THIS OLD BARN

This book is dedicated to those who truly
love old barns, especially the thousands of
Country, Farm & Ranch Living, Country Woman
and *Reminisce* readers who shared their
anecdotes and photographs. Their generosity
and deep reverence for their beloved barns
made this book a reality.

Country Books

This Old Barn
Country folks fondly recall—
in words and photos—the
heart of their homesteads.

Publisher: Roy J. Reiman
Editor: Ken Wysocky
Associate Editors: Deb Mulvey, Kristine Krueger
Editorial Assistant: Cliff Muehlenberg
Art Director: Maribeth Greinke
Design Director: Jim Sibilski
Art Associates: Bonnie Ziolecki, Tom Hunt
Production Assistants: Ellen Lloyd, Daria Mondo
Photo Coordinator: Trudi Bellin
Photo Assistant: Mary Ann Koebernik

© 1996 Reiman Publications, L.P.
5400 S. 60th St., Greendale WI 53129

Country Books
International Standard Book Number: 0-89821-175-1
Library of Congress Catalog Card Number: 96-70327
All rights reserved. Printed in the U.S.A.
Third Printing, August 2001

For additional copies of this book or information about other Reiman Publications books, calendars or magazines, write to: Country Books, P.O. Box 990, Greendale WI 53129. **Credit card orders, call toll-free 1-800/558-1013.**

PHOTO CREDITS
Cover: Tom Dietrich (historic John Moulton Barn in Grand Teton National Park, Wyoming)
Back cover: Deneve Feigh Bunde/Unicorn Stock Photos
Page 3: Bob Firth/Firth PhotoBank
Page 4: David Jensen (Wallowa Valley, Oregon)
Page 6: Richard W. Gann (near Spring Green, Wisconsin)
Page 10: G. Alan Nelson (near Muscatine, Iowa)
Page 24: Neena Mitchell-Stock/Art Images (near Winthrop, Washington)
Page 54: Bob Sisk/Dembinsky Photo Assoc.
Page 72: Darrell Gulin/Dembinsky Photo Assoc. (near Genesee, Idaho)
Page 92: A.M. Wettach/RP (Iowa)
Page 116: Bob Firth/Firth PhotoBank (Wisconsin)
Page 128: Steve Terrill
Page 140: K.B. Getz
Page 150: Tom Dietrich (near Conner, Montana)
Page 164: Bob Firth/Firth PhotoBank
Page 180: Bob Firth/Firth PhotoBank

Contents

T here's something magical about old barns. Like many folks, I can't go by one without pausing to wonder about the stories it could tell, not to mention study its ingeniously simple architecture, marvel at its craftsmanship or admire its grace and durability.

There probably are many reasons why barns tug at our emotional sleeve and urge us to take a closer look. To begin with, they're more than just a collection of beams, siding and shingles; they're bridges to the past.

Barns nostalgically remind us of a simpler time. They speak of security and warmth, of a harvest home for the winter and of beloved farm animals snug and safe from the elements.

They're potent reminders of hard yet honest, satisfying labor, of pride and resilience and of rural values we still hold dear. They evoke memories of a time when a community—in a ritual of fellowship and goodwill—would join hands to raise a barn, then hold a dance to celebrate the achievement.

The hub of a farm, barns are where our rural forefathers worked, played, danced, laughed, cried, got married, worshiped and celebrated—yes, even *lived*. As you'll read in this book, some people were actually born in a barn; in other instances, their funerals were held in barns.

Years ago, barns were so essential, in fact, that many farmers built one *before* they erected a house. If money was tight, the barn usually was the highest priority when it came to repairs or a fresh coat of paint. And on a larger level, barns often said much about a farmer's prosperity and his position in a community.

Many folks consider their "barnyard castle" a member of the family. They speak of their barns in reverent tones and compare them to churches, sometimes even referring to them as "country cathedrals". And because they were often built by fathers, grandfathers or other ancestors, folks tend to strongly associate their barns with these family members.

For the last year, we've heard all this and more from thousands of readers of *Country, Country Woman, Farm & Ranch Living* and *Reminisce* magazines who answered our call for barn photos and stories. Our request for stories was sparked by readers like Lee McKelvey of Spring Mount, Pennsylvania, who wrote and suggested that we publish a book about barns.

We were embarrassed we hadn't thought of the idea ourselves—talk about missing the obvious! You see, over the years, our mailbags have been

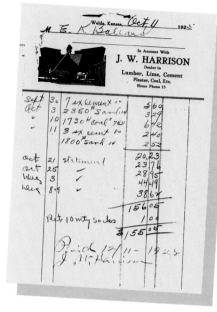

JULIA WELLING shared the original contractor's bill for her grandfather's barn, built near Welda, Kansas in 1925 for a mere $155.05!

full of stories and photos from folks expressing unabashed enthusiasm for their barns. A book about barns—hmmmm, it seemed liked a pretty fine idea.

But nothing could have prepared us for the overwhelming response, one of the largest we've ever seen here at Reiman Publications. A steady stream of mail early on turned into a deluge. For several months, we received upward of 50 letters a day. In all, we estimate more than *3,000 folks* responded. The boxes of mail filled half an office; at times, I felt like Santa Claus!

And what we found inside all those envelopes was almost as amazing as the quantity. Some readers offered to send us their entire collection of hundreds of barn photographs they'd snapped over the years. Others shared treasured vintage family photos (some over 100 years old)...personal stories of hardship and triumph...anecdotes of heroic feats of labor...and reminiscences of warm celebrations and joyful games. Some readers just were grateful for the opportunity to tell us about their family barn, regardless of whether their story was published.

Their stories reflected their deep love for their barns. "I took my first steps as a child with the barn as a reference point. Today, it's *still* my reference point," wrote Lorraine Belote from Clarksville, Arkansas. And Carolyn Hallman of Pennsburg, Pennsylvania, noted, "I know it may not look like much to you, but who I am and what I've become ties back to my grandfather Jacob Detwiler's barn (left) near Harleysville."

Julia Welling of South Bend, Indiana entrusted us with a valuable family memento—the *original contractor's bill* (above) for her grandfather's barn, built near Welda, Kansas in 1925 for $155.05. She found the receipts in a little black tin box in which her grandpa kept all his important papers.

"I PLAYED, laughed, cried, thought and sometimes even learned something in this barn—if I paid attention to Grandpa!" writes Carolyn Hallman. "The barn is gone now, but my memories will remain with me forever."

Vernon Collins of Merchantville, New Jersey sent a piece of curved wood on which he had glued a barn photo, creating a handsome three-dimensional barn that stands on my windowsill.

The extremely personal tone of these thousands of letters and other items made it awfully difficult for us to decide what would and wouldn't fit in the book; we wish we had enough pages to accommodate everything that readers were generous enough to share with us.

Because they cared enough to send us a personal piece of their lives, it strikes me that this book is a lot like the barns it honors—a storehouse filled with warm and happy memories. As I noted

BARN RAISINGS, like this one on Niles Batdorf's farm near Maple Plain, Minnesota in 1913, were community events. Niles and wife Mary are in the left half of the doorway; just to the left of them is their daughter Leona in the lap of her aunt, Martha Scheer, says Leona's daughter, Virginia Ferrin, of Plymouth.

earlier, I've often wondered about the stories barns could tell; in *This Old Barn*, they finally have a voice. I think you'll enjoy what they have to say.

—*Ken Wysocky, Editor*

Chapter 1

Big
Barns

MAINE MARVEL is an apt description for the immense Nichols Barn, which stood in Limestone and was known far and wide. It was quite an attraction; owner G. Howard Nichols used to offer guided tours for a quarter and on some Sundays would pull in $400! Grandson Warren Nichols of Caledonia, New York shared the photo.

had introduced, but what laid 'em in the troughs was the great, glistening golden cow, 8 feet high and 6 feet long, that swung majestically on the top of the cupola, nearly 100 feet from the ground."

One of Grandfather's neighbors once said that when a man got a new car, "he either wanted to see if it could climb Haystack Mountain in high gear, or he took his family to see the Howard Nichols barn."

There were so many visitors that Grandfather started to offer guided tours for a quarter. On some Sundays during summers, he collected as much as $400! His guests marveled at the tunnel, the tin-covered stable ceilings painted in various colors, and the cozy room with sofa and chairs where Grandfather could sit and admire his cows. He even had rest rooms in the basement and a lunch-counter concession.

The farmstead's glory was short-lived, though. A haymow fire in 1924 destroyed the barn, house and over 80 head of livestock. Neighbors could see the blaze from miles away, and many took off their hats in respect.

Grandfather's losses totaled $150,000, and he was insured for only $12,000. Undaunted, he moved his family into the barn cellar and later built a smaller barn on the same foundation.

—June Hechler, Solon, Maine

y grandfather, G. Howard Nichols, could neither read nor write, but his farm in Limestone, Maine was a showplace known far and wide.

In 1922, working from a model that Grandfather whittled exactly to scale, contractors built a spectacular seven-story "castle for cows" at a cost of $80,000—an astronomical sum in those days.

The barn (above) used 250,000 feet of lumber and five carloads—1,900 barrels—of cement. The 120- by 60-foot barn had 200 windows, including three huge ones made of stained glass. It was fully equipped with electricity and even had what was probably the county's first electrically controlled air-conditioning system.

A striking feature was a 150-foot-long underground passageway from the road to the barn cellar. It was wide enough to accommodate an eight-horse team and was supported by 9-foot concrete walls and 108 steel beams dipped in concrete to prevent rusting.

But Grandfather didn't stop there. He built an attached 44- by 60-foot ell, a 10- by 76-foot woodshed, a potato house, a granary, and a huge house with three bathrooms, two washrooms, a laundry room and a milk room.

"It was a great show," *The Boston Herald* wrote in 1942. "People came from all over the 48 states and beyond to gape and gasp at all the innovations Nichols

EQUESTRIAN CLASSES are held in this ornate barn at Leland Stanford Junior University in Palo Alto, California. Nancy Vynalek of Tucson, Arizona says the "Big Red Barn" and surrounding property were dedicated to Christian education in memory of the youth, who died of typhoid at age 15.

GERMAN IMMIGRANT Richard Schultz built the huge stone barn at top left on Highway 177 near Red Rock, Oklahoma. Later, it was owned by Byron and Belle Neal (that's Belle at left with granddaughters Lori and Brenda) after years of renting. Belle told daughter Patty, who shared the photos at left, that Mr. Schultz—who had no family in the U.S.—created the barn as a monument to himself. Helen Lamb of Porum took the photo above.

The largest barn in Oklahoma became part of our lives in 1947 when my father started renting a ranch near Red Rock. The owner, Richard Schultz, had completed the stone barn (above) in 1941.

Dad first saw the huge barn while hauling cattle to Oklahoma City. He was so impressed that he pulled off the road for a closer look. He remembers saying to himself, "Lord, if I could just have a place like that."

He and Mama had been moving from farm to farm since the Depression, and he dreamed of finding a place to settle down. A while later, Dad talked to Mr. Schultz about renting the property.

When they drove to the ranch to look things over, Dad said the barn was the prettiest sight he could've imagined. The barn is 160 feet long, 96 feet wide and 55 feet tall. It took 30 workers almost 3 years to build the barn from stones quarried on the ranch.

Many newspapers called the barn the largest in Oklahoma, and some said it was the second-largest in the world. (One paper later said the largest barn had burned down, making Mr. Schultz's the biggest in the world.)

The roof is made of corrugated iron, and the arched ceiling rafters are one-by-fours nailed together four boards thick. The barn has no loft, but the center section could hold 55,000 to 60,000 bales of hay. The south wing originally housed cattle, but Mama said the barn was so tight the cows got too warm and caught colds!

In 1951, a fire destroyed 50,000 bales of hay and left only the stone walls standing. Mr. Schultz rebuilt the barn 2 years later, but Dad never stored that much hay in the barn again.

In 1989, my parents bought the ranch from Mr. Schultz's heirs in Germany. But the sign on the barn still read "R. Schultz". In 1992, for Dad's 81st birthday, we surprised him with a new sign inscribed "B. Neal" (see top right photo).

While he was still alive, Dad often said the barn symbolized what farm life was all about.

"If you look at this place, the first thing you see is the barn," he said. "But look a little closer and you see good fences, clean pastures, good ponds and no trash anywhere.

"Look closer still and you can see a big family and lots of good neighbors and friends. God bless 'em, that's the heart of it all. How could I fail? I had it all. It just all goes together to make a good life."

It's hard to believe that when we first moved to the farm, the barn seemed cold and frightening. Now it represents only warmth and strength to us.
—*Patty McAlister, Clearwater, Florida*

In 1935, Walter Bones—the man who would become my father-in-law—needed a barn for bull sales at his Bones Hereford Ranch. He found one in Mitchell, South Dakota.

The all-wood barn had been built by a livestock improvement association, and Mitchell's city fathers wanted to get rid of it. Walter hired a contractor to move the building, which measured 100 feet by 200 feet.

The contractor couldn't just drive the barn down the road, so he sawed it up like pieces of a jigsaw puzzle and numbered each section. Figuring he was the only one who could put the barn back together, the contractor demanded more money when moving time came along, figuring he had Walter over a barrel.

But the contractor chose the wrong man to threaten. Walter

wouldn't give in. The barn was moved to Parker, South Dakota and pieced back together—for the original price! It was painted red with white trim and became quite an area landmark.

In 1941, the barn was destroyed in a fire. Undaunted, Walter had an exact replica built on the site and continued his annual bull sales for more than 50 years.

In the early 1960s, an electric heater started another fire, but fast action saved the main sections. Only a small part of the barn had to be rebuilt.

We no longer hold annual bull sales, but the barn still sees plenty of use. The sale ring now has a modern cattle-working system; we often use the box stalls for calving in stormy weather; and we use the pens for custom feeding. It's been painted white since those early days…and it's still quite a landmark.

—*Winnie Bones, Parker, South Dakota*

BULL SALES were held for more than 50 years in this barn on the Bones Hereford Ranch in Parker, South Dakota. The barn's round end held the sale ring, with bleachers to seat 1,500. The long section has three alleyways lined with box stalls.

VISITORS come from all over Kentucky to view the ceiling of the Richlawn Farms cattle barn in Carrollton (above), which boasts 50 curved, unspliced beams of poplar and pine (right). The barn was built in 1945 on the foundation of another barn that had burned the previous year, says Mary Ann Gentry of Carrollton. It is 203 feet long, 53 feet wide and 44 feet tall. The barn now is owned by a chemical company, which uses it for employee meetings and civic functions.

13

HUGE DAIRY BARN near Somerset, Pennsylvania bears a suggestion for thirsty passersby. Carol Thompson of Mason, Michigan spotted the majestic barn and its message while driving on the Pennsylvania Turnpike.

DRINK MILK!

THE BLOCK BARN in Cuba, New York was built in 1909 to house racehorses. It's said that a Russian czar sent his mare here to be bred. Made of cement block and poured concrete, it's an amazing 347 feet long and 50 feet wide. Gene Young of Metamora, Illinois says the barn was heated with three steam boilers and had a dance floor upstairs. It's now used for community meetings and stage productions, says LeRoy Schultz of Morgantown, West Virginia. Jean Marie Cornelius of Whitesville, New York provided the photograph.

CONSIDERED one of the biggest in Kansas, the Thomas Barn was 100 feet long, 64 feet wide and 64 feet high to the top of the cupolas. Photos by Kay Brand of Colwich, Kansas vividly show its immense size. Unfortunately, the barn burned to the ground shortly before massive renovations were to be completed.

From the moment it was completed in 1912, the Thomas Barn near Woodston, Kansas was considered one of the biggest and best in the state.

In 1918, it won a "best barn" contest sponsored by the Department of Agriculture. And as for size, the haymow alone could hold *four* ordinary barns, or an entire university-size basketball court!

Built by wheat farmer William Grant Thomas, the barn stood 100 feet long, 64 feet wide and 64 feet to the top of its cupolas, which later were destroyed in a windstorm. It could house 54 draft horses and 50 cattle. The unobstructed haymow could hold 500 tons of loose hay.

It cost $8,000 to build and required 120,000 feet of lumber, 115,000 shingles, 7,000 pounds of nails and 1,200 pounds of bolts!

In essence, the barn was a "fueling station" for the horses Thomas used to farm 1,760 acres of wheat. His grandson, Lee Hull, says, "Friends of my father can remember the parade of horses, plows and men going to the fields from the barn in the morning, coming back at noon, going back out and then coming in during the evening."

Within a few years, however, tractors made the barn all but obsolete. By the time the Rev. Richard Taylor spotted it in 1988, the roof and haymow floor were badly decayed, part of the foundation was crumbling and termites had destroyed many of the timbers.

Despite its condition, the barn was still awe-inspiring. Taylor, an engineer and history buff, felt it had to be saved.

With a small group of supporters, he formed a nonprofit corporation, Classic Big Barn Inc., to restore the barn and convert it to a museum. William Thomas' son, Clifford, donated the barn and 15 acres for the effort.

The group announced its plan in 1989, urging Kansans to pitch in. Many newspapers encouraged support for the project. Otherwise, the Wichita *Eagle* warned in an editorial, the Thomas barn would "slowly will fold its great walls and sink into the western Kansas earth".

Kansans responded in a big way, donating more than $100,000 to replace rotting timbers, give the haymow a new tongue-and-groove floor and fix the barn's foundation. The barn was listed on the National Register of Historic Places, and more than 15,000 people toured it.

Restoration was to have been completed in the summer of 1995. But that May, a fire—apparently caused by lightning—burned the barn to the ground. Laurence Hull, one of William Thomas' grandsons and president of Classic Big Barn, told a reporter, "There's been a lot of man-hours gone into it, and now it's all up in smoke. It's a hard pill to swallow."

The Topeka Capital-Journal encouraged its readers not to focus on the loss, but on the inspiring efforts of the volunteers and contributors who helped restore the barn to its original splendor.

"As one of the biggest barns in Kansas, and as a stately artifact of our history, the 'classic big barn'...was well worth saving," the newspaper said in an editorial. "If it had fallen from neglect rather than a bolt from the sky, we'd have much to regret indeed."

—Dick Taylor, Berryton, Kansas; Louise Allen, Aurora, Colorado; Marilyn Griffin, Salina, Kansas; and Kay Brand, Colwich, Kansas

"BUILT IN 1915, our barn is one of the largest in northeast Wisconsin at 170 feet long, 42 feet wide and 45 feet high," report Norbert and Rosemary Schroepfer of Bryant. "In September 1918, it was struck by lightning and burned to the ground, then was rebuilt immediately. It's been struck twice more since we purchased the farm in 1959."

GAMBREL-ROOF barn in Granger, Iowa stands 75 feet high and is 150 feet long and 50 feet wide. Greg Battles of Runnells, Iowa shared the photo. The barn is on his grandmother's farm.

WHEN BUILT in 1919, the 200-foot-long Mankato Holstein Farm Barn was the largest in Minnesota. Located north of Mankato, the barn is on the National Register of Historic Places. Photo shared by Wic Burgeson, Oceanside, California.

IMPOSING three-story horse and mule barn was the centerpiece of the Bagg Bonanza Farm in Mooreton, North Dakota in the 1930's. The 128- by 40-foot barn began to tip in 1954, lashed by 100-mph winds during a 3-day storm. The farm site is now owned by a historic preservation group, which hopes to add the barn to its list of restored buildings. Scott Holt of Lisbon, North Dakota shared the photo.

TRANSPORTING the Cooper Barn was a "moving" experience! In 1992, the mammoth barn (above) was moved 21 miles from Rexford to nearby Colby. The barn now houses farm equipment for the Thomas County Historical Society.

A longtime landmark in Rexford, Kansas, the Cooper Barn was one of the largest barns in the state. The behemoth measured 114 by 66 feet and was built in 1936 on Foster Farms for Benjamin Butler Foster of Kansas City, Missouri. Foster started farming in Thomas County in 1912 and at one time owned 33,000 acres of land—15,236 of them in Thomas County.

House siding was used to make barn more attractive. When community men finished the building, Foster hosted a dance for 2,500 people! Live music was supplied from 8 p.m. until after daybreak the following morning by two bands from radio station KMMJ in Nebraska.

To give you an idea just how big Foster Farms was, consider the 1919 wheat harvest—the farm's largest ever. It required 282 men, 24 grain binders, 27 wheat headers and eight threshing machines. In all, 64 horses were used to haul wheat bundles and pull header barges, which hauled wheat from header to stacks.

The barn housed Foster's nationally known prize-winning Honred Hereford cattle. It could house 75 head of cattle. In 1965, the farm was sold to a partnership that included a man named Willard Cooper.

The partnership dissolved in 1969 and Willard retained the land with the barn on it. He died in 1980, and relatives offered to donate barn to the county historical society—if it could be moved.

The society raised about $90,000 and planned the move for the last 2 weeks in October 1991. Later, it was postponed until mid-November to avoid heavy harvest traffic. But a major snowstorm on Halloween and wet conditions thereafter prevented any action until May 1992.

The Cooper gift contract only extended to May 31 of that year, so moving the approximately 150-ton barn took on some urgency the following spring. After much planning, the move began on May 13 after getting permission from the state, railroads, utilities, State Patrol, state Department of Transportation and Thomas County Sheriff's Department.

On moving day, it rained for first time in 3 months. But the move went on anyway. The barn lumbered along Highways 83 and 24, with American and Kansas flags flying from the ridge peak.

It was finally winched onto its new foundation at the Prairie Museum of Art and History complex in Colby at 8 p.m. on Friday, May 15. It now houses farm equipment for the Thomas County Historical Society. —*Karen Ann Bland, Gove, Kansas*

"WHEN we moved to our farm, we were told that our barn—which is 112 feet long and 56 feet wide—was the longest in the state," reports Juanita Yates of Monroe City, Missouri. "We can't verify that, but it *is* a long barn. It was a great playground for our 10 children."

ORNATE CUPOLAS and an unusual multi-paned window are unique features of this gigantic barn, which overlooks a field near Sleeping Bear Dunes National Lakeshore in Michigan. Mike Detmers of Grand Rapids sent the photo.

ALFRED VOSS FARMSTEAD, located near St. James, Minnesota, is the site of one of the largest stock barns in the state. Built in the 1890's, the mammoth barn is 180 feet long, 60 feet wide and 56 feet high.

One of the largest stock barns (above) in Minnesota stands on the Alfred R. Voss farmstead near St. James, Minnesota. Built between 1893 and 1896, it was constructed from long-lasting hardwoods that were first soaked in 2,000 gallons of linseed oil.

The barn is 180 feet long, 60 feet wide and 56 feet high. It has a gabled roof with three square cupolas and 10 gabled dormers; black walnut beams; posts of Washington fir; Washington red cedar siding; and the original wood shingle roof. It rests on a 3-foot-thick foundation that extends 10 feet into the ground.

Inside are stalls and boxes made from white oak, floors of paving brick and concrete, and a sewage and water system. It housed several hundred head of cattle and horses and could hold 1,000 tons of hay!

On one of the windowpanes in the haymow is the Latin inscription: "Palma non sine pulvare", which means "The palm is not without labor".

The old farmstead where the barn still proudly stands is listed on the National Register of Historic Places. It is currently owned and operated by the family of the late Robert Voss, Alfred's son.
—*Rodney Winter, St. James, Minnesota*

The barn on Dandelion Hill (right) in Francestown, New Hampshire was a true New England landmark. Local historians believed it was the largest attached barn in the entire region, and possibly the largest east of the Mississippi.

Dandelion Hill got its name from Ernest and Clarissa Foote, who purchased the farm in the 1930's. Dandelions flourished there, which was fine with Ernest. He loved them so much that he grew them on purpose!

But for the rest of the community, the real attraction on Dandelion Hill was the post-and-beam barn. Built in 1898 with mor-

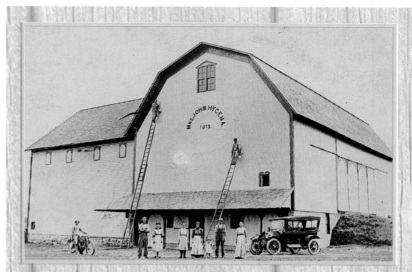

WHEN Lydia Hygema (far right) was widowed, she continued to run the family farm—with its 90- by 45-foot barn—outside Wakarusa, Indiana, says granddaughter Dorthy Hart of Marion. It was unusual then to have a woman's name on a barn. Shown are, from left: Lydia's son Elmer; son Joe; Elmer's wife, Mabel; and daughter Goldie. To Lydia's left is son Walter; his wife, Sarah; and their son, Ralph. (Men on ladders not identified.)

tised joints and wooden pegs, it stood 182 feet long and 42 feet wide, with 12-foot-square cupolas.

The entire community shared in the loss when the barn and adjoining house were destroyed in a four-alarm fire in July 1995. The blaze sparked numerous brush fires, and kept firefighters from 22 communities busy for 3 full days. Dandelion Hill will never look the same.
—*Phil Traxler, Bennington, New Hampshire*

POST-AND-BEAM BARN on the former Foote Farm in Francestown, New Hampshire was built in 1898 on a knoll known as Dandelion Hill. Destroyed by fire in 1995, it was believed to be the largest attached barn in New England.

Photos: Joel Gardner

Two huge Vermont barns believed to be among the largest ever built in the United States still stand at Shelburne Farms, a former estate that now houses a nonprofit environmental education organization. The barns, known as the Breeding Barn and Farm Barn, continue to be primary attractions at the farm, located near the community of Shelburne.

The Breeding Barn is 418 feet long, 107 feet wide and two stories high. It features a unique system of iron rods that join the side walls and the roof and holds up the roof without support beams; for 40 years, it held the largest unsupported interior space in the United States.

The Farm Barn is 200 feet long, 60 feet wide and five stories high, and has wings that are 275 feet long, 30 feet high and 28 feet wide! In addition, it included a 2-acre courtyard and a 1,500-ton-capacity hayloft.

Shelburne Farms was founded on 4,000 acres of land in the late 1880's by Dr. William Seward Webb and his wife, Lila. They wanted to create an innovative model farm featuring the most modern buildings, machinery and agricultural techniques.

Dr. Webb wanted to crossbreed English Hackneys with Vermont stock to create a horse strong enough to pull a plow, yet elegant enough to pull a carriage. Unfortunately, the internal combustion engine rendered his plans obsolete; he gave up the stock operation in the early 1900's.

In 1984, facing a huge property tax burden, the Webb family donated the farm to a nonprofit entity. Today, more than 125,000 people—including thousands of schoolchildren and families—visit the farm each year to participate in innovative environmental education programs.

The barns still stand tall and proud, a testament to Dr. Webb's ambitious vision.
—*Hilary Sunderland*
Shelburne Farms, Vermont

HEADQUARTERS for the entire Shelburne estate, the Farm Barn (top left and inset) housed offices, telephone and telegraph stations, a blacksmith shop, paint and carpenter shops, stables for workhorses and mules, fire-fighting equipment, farm machinery and rooms to clean, bag and store harvested crops. The Breeding Barn (left) was naturally lit by glazed lanterns and dormer windows. The interior, which featured a show ring, measured a whopping 375 feet by 85 feet; thousands of interior electric lights were arranged so no shadows could be cast.

19

W e just happen to have one of those great big old barns—the largest in New York when it was built in 1884. People come from all over to take tours, snap photos or paint pictures of it.

It's been called the Kennedy Dairy Farms since my husband's family bought it in 1981. But folks around here still refer to it as the "Black and White Farm", so named for the black and white animals—including mules, cows, sheep and pigs—kept by former owners Craig and Mortimore Ross.

It measures 194 by 159 and is 706 feet around. Children who played at the barn reckoned that almost seven times around the barn equaled a mile. The huge barn also is unique because it has a barnyard in the middle.

Construction began by moving three timber-framed barns to form its western side. The rest of the immense structure was built on-site.

Upkeep is the hardest pat of owning these old barns. We try very hard to keep her in good shape, as well as useful in our modern times. —*Diana Gressinger, Dansville, New York*

LARGE BARNYARD sits squarely in the middle of the immense Black and White Farm (inset), so named for the black and white animals raised by former owners. Other photos give a perspective on just how big this barn is —194 feet long by 159 feet wide and 706 feet around!

"OUR LARGE DAIRY BARN (right)—which is actually two barns connected by a third—was built in 1922 and is on the National Register of Historic Places," writes Grace Krump of Kent, Minnesota. "Located in Wilkin County, our farm used to be the second of five Femco Farms, an experiment in diversified farming begun in 1921 by Frederic Murphy. At one time, Femco Farms included 5,000 acres and had 20 of the top-producing dairy cows in the world. My husband's family moved to this farm in 1941."

"BUILT in 1849, this barn (left)—which is on the National Register of Historic Places—is 176 feet long," observes Mary Ann Miller of Wernersville, Pennsylvania. "It has a silo inside and 62 tie stalls, with a drinking bowl for each cow. The barn used to be part of a mountain health resort here on the edge of the South Mountains in Berks County. My dad, Adam Schaeffer Sr., bought the farm in 1930. Now my nephews David and Brian own it—they're the third generation to operate the farm."

When J.C. Adams built this barn (below) in Sun River, Montana in 1885, he could honestly say there wasn't another like it—at least not west of the Mississippi!

Mr. Adams hired three stonecutters to hand-cut blocks of sandstone from a quarry several miles away. The stones were hauled to the ranch on a buckboard wagon. Stones used in the archways were shipped from the Midwest by riverboat.

The barn is 140 feet long and 40 feet wide. The main floor had stalls for racehorses and space for wagons, buggies and coaches, and one corner was used for ice storage and a beef cooler. The hayloft was the scene of numerous gatherings.

In later years, the barn's lower section was used as a dairy barn. Now vacant, it was recently restored and is listed on the National Register of Historic Places.

—*Ruth Nina Merja*
Sun River, Montana

AT ONE TIME, this was the largest mule barn in Kentucky, reports Max Soaper, grandson of William Soaper, who had the barn built in 1926. The barn, located on the outskirts of Henderson, Kentucky, is 168 feet wide, 40 feet wide and two stories high. It burned to the ground when it was three-quarters finished, but William persevered and rebuilt it in 1927. It boasted 30 stalls and a hayloft spanning its entire length and breadth! Max's friend Betty Smithhart took the photo.

STONE BARN in Montana features unusual Romanesque arches over some of the doors and windows. To get some idea of its size, consider this: The stallion weather vane is life-size!

My grandfather, George Souers, had a thriving horse-selling business in Huntington, Indiana at the turn of the century. He initially bought horses from U.S. distributors, but in 1906, he began traveling to Europe to buy horses directly from breeders. His Belgian, Percheron and German coach horses—said to be among the finest ever seen in Indiana—could bring up to $4,000 each. But first they had to be trained, and these enormous animals required lots of space and feed.

The family barn just wasn't big enough, so in 1908, my grandfather built a new one. The local newspaper called the barn (below) "a masterpiece in engineering and planning". It was three stories tall, 60 feet wide and 150 feet long, with stalls and floors of reinforced concrete.

Each stall had running water, and the systematic feeding system could boost a horse's weight by 100 pounds a week—but remember, these were *big* animals. The biggest, the Belgians, weighed about 2,300 pounds.

At the height of the business, my father and grandfather went to Europe on buying trips two or three times a year. But the rise of the tractor and the start of World War I—which made ocean crossings impossible—ended all that.

The business closed in 1916, and the barn was sold to a rubber company. Although the barn lives on as a manufacturing facility, from the outside it still looks much as it did nearly a century ago.
—*Max Souers, Vancouver, Washington*

ONE OF the largest barns ever built in North America stood near Leader, Saskatchewan. Photo was sent by Leader resident Brenda Flood.

One of the most famous ranchers in Canadian history had a reputation for doing everything in a big way, and his legendary barn was no exception. In fact, some believe it was the largest barn ever built in North America.

William Theodore Smith, a Kentucky native, homesteaded along the Saskatchewan River in 1868. By 1914, his operation had grown to 10,000 acres, with more than 13,000 head of livestock and a huge alfalfa operation. Smith needed a *big* barn. And he got one.

The impressive structure was 400 feet long, 128 feet wide and 60 feet high, with seven ventilators and numerous windows. Smith financed the construction with a loan of about $82,000, which he took home from the bank in a gunnysack.

Construction required 32 train cars of lumber and 30,000 bags of cement. When Smith ordered an entire train car of nails for the project, the wholesaler thought he was joking and didn't even fill the order!

One hundred laborers and craftsmen worked for 5 months to finish the barn. Then Smith hosted a barn dance unlike any other, with guests from as far away as Montana and California. Two orchestras provided music in the hayloft, which was nearly a city block long.

Smith fell ill shortly after the barn was completed and died in 1918 at age 73. In 1921, a trust company took over the ranch. The renowned barn was dismantled and the lumber sold to local farmers, who used it to build their own much smaller barns and sheds.
—*Menno Kliewer*
Reedley, California

GEORGE SOUERS BARN in Huntington, Indiana is still a town landmark. Photograph was shared by Sandi Tucker, great-granddaughter of George Souers. (The sign looks backward because the photo was taken from the rear of the barn.) Sandi lives in Coolville, Ohio.

"MY great-grandfather, Jasper Davis, rented out his land for a year so he could concentrate on building this cattle barn southeast of Williamsburg," explains Paula Sue Smith of Williamsburg, Kansas. "It's 185 feet long and two stories high. During WWII, bomber pilots practiced dives by aiming at the barn because it was the biggest 'target' in the area. We used to have basketball tournaments in the hayloft!" Above photo shows barn and Jasper and family in 1890; at right is the barn today.

THIS BARN (above) outside Otranto, Iowa, near the Minnesota border, was 64 feet wide and 96 feet long. "It was built in 1896 by a neighbor of ours, Pete Hanson, if I remember correctly," writes Roger Walsh of Austin, Minnesota. "The man sitting on the roof is Russ Lee, who owned the farm after Pete. In this photo, taken in 1936, the barn was being reroofed; it took more than a boxcar load of shingles to do the job."

TILE SKIRT dresses up this barn (above) along Highway 10 near Chili, Wisconsin, notes Marion Johnson of Greenville, who snapped the photo. "My dad built the main structure in the 1930's and added on the round-roof section in 1967," explains James Gilbertson, who operates the farm for his parents, Virginia and Vernon. "The tile was trucked in from Iowa, and Mom painted the cows on the side." The barn is 114 feet long!

REPORTEDLY the largest barn ever built in Minnesota, this beauty was referred to as the "Big Barn" when it was built near Grasston for farmer John Runquist in 1907, writes Shirley Hansen of Grasston. The barn was so big—75 feet wide, 125 feet long and 79 feet high—that grain could be threshed within its walls! It burned to the ground in 1963.

Roadside
Art-tractions

Chapter 2

Wood Canvases

Sometimes, a red barn just isn't enough. To be sure, few things are more eye-pleasing than a classic red barn. (In fact, did you know Northern farmers started the trend by mixing iron oxide, skim milk and lime to create a paint-like preservative that was decidedly red?)

But to some folks, all that wood is nothing more than a big country canvas just begging to be filled. So, sit back and "brush up" on the following gallery of fine barn art. We're sure you won't be seeing red...

SOARING EAGLE on patriotic barn in Rosemount, Minnesota caught the eye of Connie Cleaver, Nampa, Idaho.

STUDENTS from Fort Atkinson, Wisconsin painted the eye-catching mural above for the nation's bicentennial in 1976. The barn's owner, the late Hilbert Schneider of Johnson Creek, was so proud of his "Blue Cow" that he even installed floodlights for night viewing! During 1976, it was featured on the NBC evening news and on the cover of *Time* magazine. The local landmark was destroyed by a tornado in 1980. Don Condon of Janesville shared the photo.

BARN along Interstate 101 south of Gilroy, California is a big hit, says Bea DeLong of Lake Albano. Mural is the work of artist John Cerney of Carmel. Notice the three-dimensional effect created by cutout figures.

25

s a child, I longed for crayons and watercolors—things I couldn't have—but instead drew and sketched with pencils.

Then I saw an advertisement for a patent medicine painted on a barn. The picture was so large that it made me want to draw bigger and bigger sketches—but where?

Time passed, but the urge to paint and sketch remained. I married, had three children and became a grandmother and great-grandmother before I found the time to paint. I retired from teaching and began taking painting lessons, which gave me a new appreciation for our 100-year-old farm near Hildreth, Nebraska.

And suddenly I knew how to fulfill my lifelong urge to paint something big. Our barn would be the canvas! It needed plenty of work—new shingles and siding, window and door repairs and several coats of paint. But when we finished, I could paint my scene there.

One summer we started fixing up the barn. The minute we finished, I took one look at the shiny coat of fresh white paint and saw the entire mural in my mind. Now all I had to do was paint it!

I started by painting a horse that is almost life-size. Our tenant spotted it from about a quarter mile away as she approached the driveway and thought someone's horse had strayed into the yard. When she opened the car door, she said, "You sure fooled me. I thought that was a real one!"

For the next 5 days, time vanished as I painted, constantly climbing up and down a ladder and stepping back to make sure everything looked right. It wasn't easy—at over 70 years old, I wasn't accustomed to all that climbing! But with God's guidance

FOR YEARS, Lillie Bunger dreamed about painting a barn scene. She finally did it in 1989—when she was over 70 years old! The family farmstead is located on Highway 4 near Hildreth, Nebraska. The mural, which Lillie named *Recollections*, is about 42 feet long and 20 feet high.

and help, my dream finally became a reality.

The scene reflects our lives in Nebraska and Washington. It shows a horse awaiting a rider—our son across the fence in the pasture with his dog, hunting squirrels and rabbits...our daughters holding hands, walking toward the tire swing...haying time...cattle heading for the water tank...the mountains near our Washington home...and husband Elmer and me sitting on a bench, a kitten at our feet, recalling and savoring it all.

—Lillie Bunger, *Woodland, Washington*

COCKEYED COWS were spotted in Canton, New York by Janice Risley-Smith of neighboring Corning. The owners told her their son had painted the whimsical duo the year he graduated from high school. The young man has since completed college and now has a farm of his own.

AS A CHRISTMAS PRESENT in 1994, Mark Nolan painted this 30-foot bull on Dad Tom's tobacco barn. Mark, who owns a sign-painting firm, got a helping hand from his brother and brother-in-law. "We had a ball and laughed every time we ran back from the barn to see our work as it progressed," he recalls. The Nolans live in Danville, Kentucky.

PATRIOTIC BARN is located on Highway 72 between Empire and Traverse City, Michigan. Gene and Marie Peiter of Longview, Texas took the photo in July 1988.

WHEN they say they've got a "big head of cattle" at the Musgrave Brothers farm in Monrovia, Indiana, they're not kidding! Betty Musgrave based her enormous painting (at right)—it's 27 feet wide and 18 feet tall—on sketches of cattle on the farm. The painting took 8 weeks to complete. Then Betty painted the opposite end of the barn with a cow's backside as a practical joke!

MUSIC LOVER must have envisioned this massive violin, complete with bow, that appears on a silo along Highway 52, south of Dubuque, Iowa. Joan Brown of Marion shared the photo.

CIRCUS THEME is featured on brightly colored barn near Janesville, Wisconsin. LeRoy Schultz of Morgantown, West Virginia shared photo.

FOUR SEASONS mural was part of the package when Wava Nelson bought property just outside McPherson, Kansas. "I was told it was painted by students from McPherson College," Wava recalls. "When the delightful paintings faded, I had the barn painted red."

A PAIR of huge paintings—"Black Baldy" on the barn (right) and "Loch of Scotland" on the machine shed (above)—brighten Paul Forsyth's farm outside Franklin, Minnesota. The larger mural depicts the house near Aberdeen, Scotland where Paul's grandfather was born, and includes a Scottish seascape and bagpipe player. Both were painted by Gary Butzer of Morton, Minnesota.

A WISCONSIN Arts Board project has restored barn paintings like this Native American portrait, says Jackie Wagner of Schofield. Another program restored a badly weathered mural on the side of this barn. Albert Brunsch owns the barn, which is on Highway 51, just south of Merrill.

IT MAY not have a mural, but have you ever seen a barn painted green before? When Georgia McGee of Willard, Ohio saw this one in Amish Country, near Sugarcreek, she made her brother-in-law back up so she could take a quick picture. She liked the photo so much she had it turned into a jigsaw puzzle for her father.

MONA LISA'S face is as inscrutable as ever here, but the sentiments on her T-shirt are perfectly clear. The owners of this barn on Highway 64 near Cornell, Wisconsin are clearly University of Wisconsin fans—and maybe hunting enthusiasts, too (notice Mona's camouflage jacket!). Laura McKeag of Grantsburg, Wisconsin shared the photo.

AN AVID Beatles fan, Carolyn Terrell's son stopped to meet the owners of this barn outside Thompson, Ohio and quickly became friends. The mural was created by the owners' daughter. Carolyn lives in Painesville, Ohio.

CONTENTED HERD of Holsteins gazes out from the side of this barn near Nappanee, Indiana. LeRoy Schultz of Morgantown, West Virginia provided the photo.

REMEMBER barn ads for Bull Durham tobacco? One was painted on Ken Lore's barn in Newport, New Jersey in the early 1920's, but it disappeared over the years. With the tobacco company's consent, Emily Stites, also of Newport, drew the bull to scale and then made stencils so she and Ken could re-create the image. "It's not exactly like the original, but he appears to be a spirited animal," Emily observes.

LONGHORNS loom on the side of this barn on the Gravett family ranch near Catlett, South Carolina. Mural was painted by Sylvia Lindemann of Pawleys Island.

REALIZING a 20-year-old dream, Blain Blakeslee of Union City, Pennsylvania painted a mural of his Uncle Richard Butler's barn right on the barn!

MURALS—and a bull—are busting out all over at Rocco and Linda Manno's dairy farm in Warwick, New York. The artist is son Rocco II, whose first painting was a 10-foot rendering of a Normal Rockwell painting. That effort drew so many compliments that he decided to add a 13-foot bull jumping through the barn wall. He's also painted a horse, cow and cat on the front of a shed. His next project? Painting a mural on a 40-foot silo!

COWS "coming home" near Madison, Wisconsin caught the eye of Jim Scurlock of Madison.

ANIMAL SCENE was painted by Debbie Van Vleet on her and husband Jeff's farm near St. Johns, Michigan, reports mother-in-law Shirley of Higgins Lake. "A friend of Jeff's did the outlines and Debbie painted them," she notes. "People come from all around to see their work."

MESMERIZING LIKENESS of Leonardo da Vinci's Mona Lisa makes this barn near Plymouth, Michigan a work of art. Photo's from LeRoy Schultz of Morgantown, West Virginia.

THE "MOUSE BARN", as it's known around Whitsett, North Carolina, has animated Tom Kleeberg's farm since 1971. In 1976, Tom gave the 30-foot mouse a bicentennial makeover, painting an American flag and a cake that says "Happy Birthday, America".

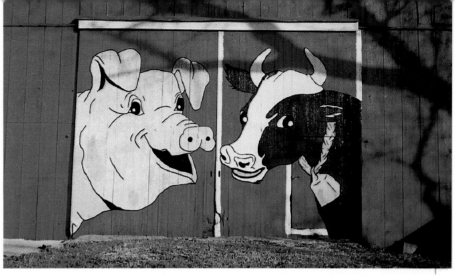

FRIENDLY CRITTERS decorate the barn on the farm owned by John and Suzanne Forsythe. Daughter-in-law Cindy of Nescopeck, Pennsylvania sent the photo.

SYMBOLS representing Dodge County, Wisconsin are painted on this barn on Highway 151, between Columbus and Beaver Dam. Don Condon of Janesville took the photo.

CRITTER DUET was spotted by Marjorie Scott of Broad Run, Virginia in Prince William County.

OLD MASTER portrait was snapped in Livingston County, Michigan by Marian Felmlee of Ann Arbor.

WHEN curious passersby ask about the mural on their Shortsville, New York barn, Mr. and Mrs. Thomas Mattice never hesitate with an answer—the artists were their 13- and 11-year-old grandsons! The project, which was Grandma's idea, took 3 days to finish, says Marie Butler of nearby Macedon, who shot the photos.

REPRODUCTION of Grant Wood painting adorns the barn door on Ginny Speeter-Wippel's farm in Paw Paw, Michigan. Ginny's American Gothic painting is 8 feet high and 13 feet wide.

THIS wagon shed/garage is near Norm Blodgett's home in Haddam, Connecticut. He provided the photo of this appealing farm scene.

CUTOUTS on this farm in Newton, Wisconsin immortalize Brian Goedtke's three daughters doing their chores. Brian took pictures of the girls as they worked around the farm, then enlarged the photos, cut the silhouettes out of plywood and painted them, says Kermit Hackmann of Manitowoc, Wisconsin.

PEACEFUL summer scene was painted by Carla Arneson of Alexandria, Minnesota during a summer when she worked at Sunnybrook Farm, a camp for handicapped kids located in Mound. "The owner asked me to paint a mural on the end of the barn," she recalls. "I'm not fond of heights, but I did it anyway. When I look at this picture, I'm glad I did—what an experience!"

RANDALL KOPF always wanted a place in the country —and a white buffalo on the barn. His eldest daughter painted this mural, based on a 1910 White Buffalo cigar box, at the Kopf home in Lexington, Nebraska. "Our barn can be seen from the road and people usually slow down to look as they drive by," notes Randall's wife, Nancy.

BARN near Lancaster, Pennsylvania has been converted for use as a play center for children, says LeRoy Schultz of Morgantown, West Virginia. The quote painted on it reads: "Judge a nation by the smiles, by the laughter of her children."

EACH SPRING since the late 1970's, high school seniors in Franklin, Virginia have descended on this barn to paint a "college collage" of their names and those of the schools they plan to attend. Photographer James Vasoti of Franklin says the owner has only one condition for this annual ritual—at least one graduate has to be headed for the Virginia Military Institute!

isconsin farmer Gene Schuster didn't think much of it when he mentioned to his daughter, Linda, that the family barn needed painting.

And he certainly didn't know what to expect when she offered to paint it for him—provided she could do it her way.

Now, thanks to Linda's imagination, the Schuster barn on Highway 19 is a local landmark here in southern Wisconsin. I should know—I live only a half mile away from this roadside masterpiece!

Linda spent about a month designing the now-famous murals. She and a friend spent another 2 weeks scraping and priming the buildings. Then came the actual artistic part, which required 15 gallons of paint and 3 weeks to complete, working in between farm chores.

The design is filled with things both whimsical and real. Linda started with a huge rainbow to brighten up the place, then added a hungry-looking combine for some more comic relief. If you look closely, you can see her brother, Dave, resting under an oak tree painted on the front of the building adjoining the barn.

Linda also decided a tractor would be nice, so she checked with local farmers to see which brand they preferred. Those she polled were split evenly between two brands, so she used both to avoid disappointing anyone. The two tractors meet at the corner of the outbuilding, where they create a giant smiling face. Linda also added dancing cornstalks and other local images.

Last but not least is the dairy cow that adorns the front of the barn. As you can see, she's quite a star in her own right.

RAINBOWS, hoofprints and a cosmopolitan cow are among the elements Gene Schuster's daughter, Linda, incorporated in her whimsical barn mural (right). Outbuilding (below) features a ravenous combine, dancing cornstalks and a hybrid tractor with a smiling face.

Linda completed the initial artwork in 1980, and repainted the barn in 1992. That's great news for people who want to take pictures. Just hook up to Highway 19 north of Madison and head east. Where's the Schuster farm? You'll know when you get there!

—*Suzette Buhr, Marshall, Wisconsin*

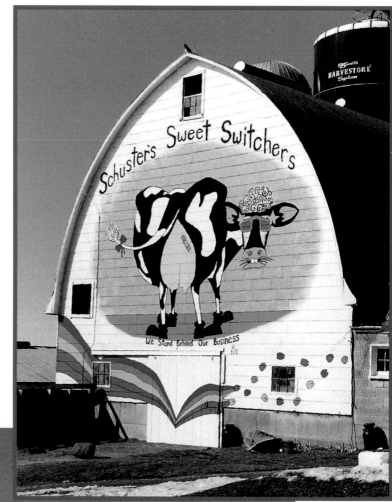

CALF BARN at Kankakee, Illinois provided a nice canvas for artist Michelle Hove (far right) to display her artistic talents. From start (above—that's Michelle kneeling and author's daughter-in-law Jeanette standing) to finish (above right), it's a masterpiece! Note detail in photo at right.

About 16 years ago, we asked a family friend, high school student Michelle Hove, to paint a mural of our sons' Holstein show cattle on our calf barn.

Since then, we've become grain farmers and decided it was time to replace the mural. We again called upon Michelle, now a free-lance artist in Chicago, to paint a harvest scene. She worked from photos we'd taken during harvest plus pictures of farm machinery from our grandson's scrapbook.

Michelle included every aspect of our farming operation, past and present—the big green machinery, Holstein cattle, Pioneer seed and our semi. The trompe l'oeil ("fool the eye") style makes the scene look as if it's in motion from every angle.

My husband, Bob, and I were dairy farmers for many years. Now our son David and his family live on the farm and enjoy the mural all year long. They keep it lit at night, so regardless of the hour, it's magnificent!
—*Joan Butz, Kankakee, Illinois*

WHO loves horses? James Shelton does—and his barn (above) shows it! When he learned of an artist who'd painted a horse on a barn roof in Texas, James contacted her for a similar project at his place south of Houston, Missouri. James added the finishing touches, decorating one side of the barn with horseshoes.

FARMING TRADITIONS are reflected in this mural on family farm near Fulton, Kentucky, says Jean Burnette. "Our farm used to belong to my husband Jeff's Great-Uncle Cecil. We were restoring the house and barn when I got the idea to paint a mural on the end that's seen by motorists as they travel the highway in front of our house. Jeff's dad, granddad and great-granddad all farmed and his grandfather was a John Deere equipment dealer. My brother, Richard Pickle, did most of the painting; I did the trim and the background."

HORSIN' AROUND is a common occurrence at Trio Farms Horses near North Manchester, Indiana.

DEER live year-round on the Whitney Barn located near Warsaw, Indiana.

NO NEED to wonder what this Indiana farmer raises—the answer is up on the roof!

THE CALLOWAYS

HOLSTEIN makes visible statement on the Schipper Barn near Roann, Indiana.

DESIGN is Deere to farmer's heart on the Hosler Barn near Columbia City, Indiana.

Look Up This Art!

Barn art seems to be taken to new heights in northeastern Indiana, judging from the photos on these pages, snapped by rural photographers Ted Rose and Jerry Irwin (lower left photo only). These rooftop murals are among more than 200 designed shingle-handedly by roofer Tom Rentschler of Rochester. "It's a way for farmers to make a statement," he says. "I can't believe how popular they've become …many of them are local landmarks."

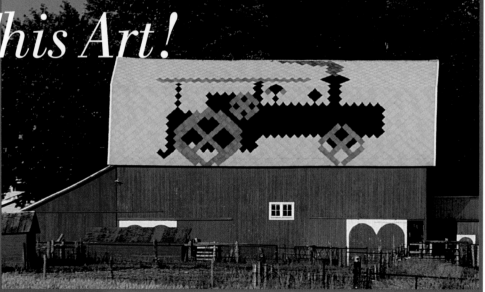

"OFF-ROAD VEHICLE" is prominently featured on this barn's roof outside Pierceton, Indiana.

YOU CAN'T DUCK the design on the McDonald Farm at Silver Lake, Indiana.

Barn Ads

Before billboards and television, sturdy barns located within eyeshot of country roads were the perfect vehicles for outdoor advertising.

But the growth of radio and television, the development of interstate highways and the federal Highway Beautification Act (which banned commercial signs within 660 feet of federally funded highways) changed all of that. Today, many barn ads are nothing more than waning relics.

Despite their fading condition, though, these slices of rural history are as alluring as ever, as you can see from these photos and those on the following pages.

PARK RANGER Bob Davis painted the sky on this barn, then outlined and numbered the rest for volunteers to finish. The barn is at the north entrance of the Hollister Hills State Vehicular Recreation Area, which is part of the California Parks Department.

"MY CLASSIC BARN with attached cow shed has two billboards, circa 1800, that advertise a clothing chain and a shoe store," explains Russell Hartz of Monroe, Connecticut. "Both buildings are in fine shape and are frequented by many artists and photographers. I did research on the billboards and had them restored."

When the State Vehicular Recreation Area at Hollister Hills, California expanded, an 80-year-old barn came with the property. The barn stands at the park's north entrance, and we wanted to make it a landmark to welcome our visitors.

We contacted several artists about painting something memorable, but the cost was too high. After much head-scratching, we turned to one of our own—Ranger Bob Davis.

Bob had never tried such a large painting project, but was willing to try. He spent weeks painting the sky and outlining the hills and wording, then let volunteers take over.

Armed with color-coded paint cans, more than 25 aspiring artists scampered up and down scaffolds all day, filling in the areas Bob had pre-drawn and numbered. When all was said and done, they were pleased with their work—and the park had an unforgettable entrance. —*District Superintendent Ralph Fairfield*
Hollister, California

HUNGRY DRIVERS get a clue regarding their next meal ahead, thanks to this colorful ad on a barn off Interstate Highway 65 in Indiana. Jack Westhead took the photo.

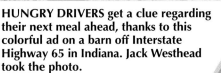

DISTINCTIVE LOGO decorates side of barn near Brooks in Oregon's Jefferson County. Barry Peril of Portland, Oregon took the photo.

CAR DEALER gets good exposure with ad on the end of this barn on the Diamond M Farms on Highway 2, between Snohomish and Monroe, Washington. Photo was taken by Ruth Abbott of Delta Junction, Alaska; her two girls are second and fourth from left in photo, while the other two children are friends whose father managed the farm.

A CLEVER SLOGAN—"Come to church to get your horse shod"—marked the blacksmith shop Emmett Church operated in his Middleton, Idaho barn. His granddaughter, Barbara Shields of Yorba Linda, California, says he painted the words and pictures himself. The photo was taken sometime between 1907 and 1910.

OLD BARN on Highway 14, south of Richland Center, Wisconsin, advertises a brand of flour. LeRoy Schultz of Morgantown, West Virginia shared the photo.

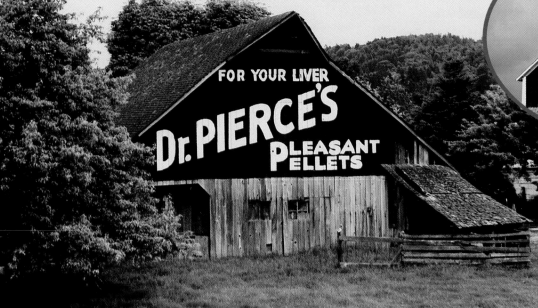

FOR YOUR LIVER
Dr. PIERCE'S
PLEASANT PELLETS

A GENERAL TONIC
Dr. PIERCE'S
GOLDEN MEDICAL DISCOVERY

ANOTHER "Dr. Pierce" barn ad stands in restored splendor near Kalispell, Montana. Nick Haren of Kalispell submitted the photo.

LANDMARK "Dr. Pierce Barn", a prairie or Western-style structure, was built around 1900 in Oregon's Willamette Valley. Thanks to contributions from barn enthusiasts, it was restored in 1990 with siding salvaged from other barns. The patent medicine advertised was a popular remedy in the late 19th and early 20th centuries.

The "Dr. Pierce Barn" has been a part of Cottage Grove since about 1900 and features a rare example of turn-of-the-century advertising for "Dr. Pierce's Pleasant Pellets".

The Cottage Grove Historical Society talked to the owner several times about buying the barn and preserving it, to no avail. Then one day in 1989, we learned the property had been sold—and that the buyer planned to tear the barn down! The thought was more than we could stand.

After some negotiating, the new owner granted us a 20-year easement, and we agreed to restore the barn. A "Save the Barn" committee mailed hundreds of fund-raising letters, held yard and garage sales and sold commemorative buttons. The contractor agreed to take payment as contributions came in.

And they did come in—from residents of other states, members of other historical societies and former Oregonians who fondly remembered the barn.

Restoration began in April 1990. Delighted citizens called every day to say they'd noticed the roof was being fixed, or just to say how great the barn looked. We were amazed at how many lives this restoration touched.

The project cost just under $12,000. When it was finished, we held a dedication ceremony in a parking lot across the highway, with a full view of the barn and sheep grazing in the pasture. The barn is now on the National Register of Historic Places. As far as we know, it's the only "advertising barn" listed.

To pay off the remaining debt on the project, we sold reprints of Dr. Pierce's First Aid Book, which someone found in an antique store. Until then, we didn't even know Dr. Pierce was a real person! That discovery made the barn even more special.

There are still a few barns with painted ads in the Western states, but most are about to fall down or have signs that are nearly illegible. We feel fortunate to have been able to save our barn—and we hope it lasts for a long, long time. —*Marcia Allen*
Cottage Grove, Oregon

CHIEF OSHKOSH BEER

ON THE WAY to vacations in Door County, Wisconsin, Cheryl Sanders recalls that her family always watched for this familiar landmark. Cheryl lives in Waupun, Wisconsin.

ALTHOUGH weathered and faded, the Miller High Life sign on this barn is still clearly visible. Patty Reynolds of Saginaw, Michigan discovered it in Boyne Mountain.

CLASSIC gable-roof barn featuring fading ad for tire recapping company was built before the turn of the century, says LeRoy Schultz of Morgantown, West Virginia. The barn is located near Lake Delton, Wisconsin.

"WHITE AND RED BARNS are common, but I think black barns are unusual," writes Freeda Haggard of Redding, California. "We saw this one while driving through Ohio."

THE SILO next to this gambrel-roof dairy barn near St. Peter, Minnesota sports a modern soft-drink logo. The photo was submitted by Gordon and Lois Robinson of Minneapolis, Minnesota.

SPIRITUAL MESSAGE tops this barn located between Freeport and Rockford, Illinois on Route 20. "It inspires me to say, 'Thank you, Lord,' every time I pass it," says Theresa St. Ores of Freeport, who shared the photo.

A barn on Highway 20 between Freeport and Rockford, Illinois has uplifted many spirits over the years—and even saved a life.

The story begins years ago, when a local church painted the fence posts on Gilbert Crull's farm with a series of religious messages, similar to the then-popular Burma-Shave signs. Over the years, the signs deteriorated badly and needed repair.

Gilbert and his sons discussed that problem—and the barn roof, which needed new shingles. Rather than fix the signs, they decided to shingle one large inspirational message into the roof. They considered "Repent" and "Jesus Saves", but settled on "To God Be the Glory", the name of one of their favorite hymns.

Eldest son Steve measured the roof and drew a blueprint for the shingle design while away at college. That spring, Gilbert and his sons carefully numbered the shingles—dark green for the background, white for the letters—and put them in place. Neighbors referred to the building as "Mr. Crull's Glory Barn".

One day the Crulls found a note from a young girl in their mailbox. Her boyfriend had died in a motorcycle accident, and she was so despondent that she'd planned to drive her car into the path of an oncoming semi. "I saw the sign," she wrote, "and decided not to do it."

—Mrs. Merrill Welling, Rock City, Illinois

BARN near Aberdeen, South Dakota is definitely on the sunny side of the street, as sign attests. "The folks who owned our farm before us called the place 'Sunny-Side', and we do, too," says Carol Lowe, who sent the photo.

LOCATED in Nevada City, California, the Celio Barn was named after William Celio, a prominent and enterprising merchant who arrived in northern California in 1881, reports Gaylo Conner of Grass Valley, California. In 1909, William became involved in the grocery business and used this barn across the street from his store for storing hay and groceries. Today, it serves as a resting place for horses that pull old-fashioned carriages through the city's streets.

KOEHLER BREWERY of Erie, Pennsylvania no longer exists, but this sign touting the "Koehler Collar" still greets travelers on Highway 5, 10 miles west of the city. Ads on each end of the barn have existed since at least the early 1950's and were recently repainted, says Don Bajorek of Fairview, who snapped this photo. Other ads for the once-popular local brew featured a burly Jackson Koehler hoisting a frosty mug.

OLD BEECH-NUT BARN was spotted south of Red Wing, Minnesota by Ron Hendrix of Apple Valley.

POTATO CHIP AD on the side of this silo near Portland, Michigan is still legible, says Dennis Simon of Portland. His uncle, George Smith, took the picture.

TOURIST ATTRACTION gets a plug via this barn roof in Coffee County, Tennessee. Photo was snapped by Robin Rudd.

COCA-COLA goes good with this white barn on U.S. Highway 3 near Mt. Vernon, Ohio. Jack Westhead of Plainfield, Indiana took the photo.

MORE than one ad suits this weather-beaten barn in Bunker Hill, Ohio just fine. Jim Witkowski of Akron, New York shared the photo.

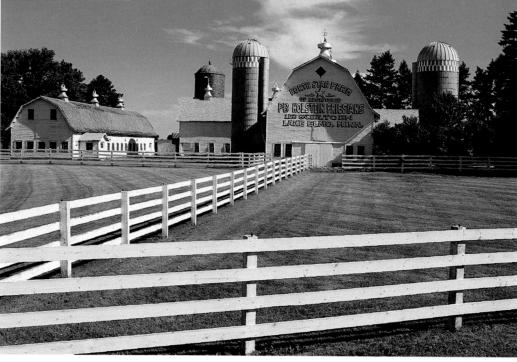

BRIGHTLY PAINTED SIGN on a livestock barn offers directions to a local church and a welcome to "God's Country" near Fair Ridge, Pennsylvania. LeRoy Schultz of Morgantown, West Virginia provided the photo.

HERD of registered Holsteins on North Star Farm near Lake Elmo, Minnesota gets plenty of notice, thanks to this barn. Richard Gann of Eaton, Colorado took the photo.

THE MAY COMPANY, based in Cleveland, Ohio, owned a chain of department stores until another company bought it in the early 1990's. Since this ad referred to the May Company's growth, it must have been painted many years ago, observes Ken Federer of Parma, Ohio. The barn is in Middlefield, Ohio.

As Depression-weary Americans rediscovered their love of the automobile in the mid-1930's, two popular tourist attractions found a surefire way to capture their attention—barn billboards.

Rock City Gardens fired the first salvo in 1935, when owner Garnet Carter hired self-taught painter Clark Byers to swab a catchy phrase on a barn near Kimball, Tennessee. That was only the beginning.

Over the next 33 years, Byers traveled from the Great Lakes to the Gulf of Mexico, looking for the most visible barns for Rock City's distinctive white-on-black slogans. In their heyday, catch phrases like "See 7 States From Rock City" appeared on about 900 barns in 19 states.

In addition to painting and dreaming up slogans, Byers negotiated deals with the barn owners. They usually got free passes to Rock City, an armload of promotional items and sometimes the modest sum of $3.

Meanwhile, in Stanton, Missouri, entrepreneur Lester Dill was promoting Meramec Caverns on historic Route 66. While traveling in Virginia and Tennessee, he spotted ads painted on barn roofs.

Lester liked the idea so much that he started a barn ad campaign of his own. Like Rock City's ads, his featured large white letters against dark backgrounds.

Ads for Meramec Caverns (like the one below) soon were as common in the Midwest as Rock City's were in the Southeast. In some "border" states, like Illinois, signs for both attractions could be found within miles of each other!

Today, there are fewer than 100 Rock City barn ads. Many were removed or painted over in the mid-1960's after federal legislation curbed roadside advertising. A painting crew still touch-

PAINTER Clark Byers posed for this photo of his first Rock City barn sign near Kimball, Tennessee in 1935. He painted all his signs freehand and admits to making a few mistakes along the way. "But most of the time I got it right," he notes. Now retired and living on a farm in Georgia, Clark still writes to some of the folks he met during his "barnstorming" days.

es up the handful of remaining signs.

While still open to visitors, Meramec Caverns no longer maintains its signs, so many are barely legible today. But the few that remain offer a faint reminder of a gentler time, when "advertising wars" were waged on the roofs and walls of humble barns.

—*Tim Hollis, Dora, Alabama*

CLASSIC Meramec Caverns ad stood out in full color on this barn near Plainfield in Hendricks County, Indiana. Jack Westhead of Plainfield took the shot.

THIS BARN, located on Highway 72 between Stevenson and Scottsboro, Alabama, was snapped by Juanita Hookey of South Pittsburg, Tennessee.

ROCK CITY BARN on Highway 64 west of Lawrenceburg, Tennessee was photographed by Roger Black of Hermitage. Rock City ads are one of Roger's favorite subjects (also see photo at bottom).

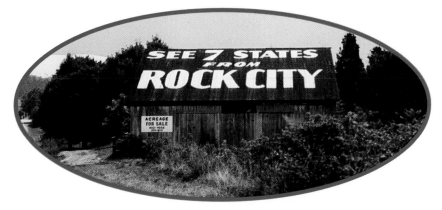

STILL-SHARP AD is visible to drivers on a well-traveled highway in Seymour, Tennessee, says Glen Barnes of Knoxville, who took the photo.

ROGER BLACK has photographed many Rock City ads over the years but has seen only one with this slogan. The full message on the tin roof originally read, "See Incomparable Rock City". The barn is on Highway 41A, south of Clarksville, Tennessee.

NEAR Dalton, Georgia, Carolyn Sue Connerly of Holly Hill, Florida pulled off the interstate and hiked into a field to photograph this barn.

MOST ROCK CITY slogans were painted on barns' metal roofs, but Jeanette Fortenberry of Kennesaw, Georgia found a neatly painted message on the side of this well-kept Tennessee barn.

47

I magine painting the same message again and again, for 50 years, on some 20,000 barns. That's the rustic legacy of Harley Warrick. Until his death in November 2000, Harley had gained legendary status as the last Mail Pouch barn painter in America.

"It takes me about 4 hours to paint a sign from scratch," Harley said back in the mid-1990s. "You can't go through these things like Grant through Richmond."

Harley painted the letters freehand—he'd done it so many times, he knew every stroke by heart. "After the first 10,000, you get the hang of it," he once joked.

Mail Pouch ads began appearing on barns in the Midwest and South during the Great Depression. For cash-strapped farmers, the offer of a free paint job was just too good to pass up.

"Farmers were happy to see us," Harley recalled. "There was no money for improvements, so this was a way for them to get at least part of their barns painted for free.

"And if you were lucky enough to live on a well-traveled route, the company would send you a small stipend for letting them use your barn."

At one time, Mail Pouch had four crews, each painting three or four barns a day. Harley's territory included West Virginia, Pennsylvania, Maryland, Ohio and Indiana.

As laws were passed to limit roadside advertising, the painting crews were phased out until only Harley was left. For many years before his retirement, Harley was the only Mail Pouch painter in the nation.

When it had a full complement of painters, Mail Pouch touched up each barn every 3 to 5 years. Today, farmers must maintain the signs themselves, or watch them slowly fade into memory.

"It's a part of American history," Harley reflected. "You hate to see 'em go."

That's why, from time to time, he'd still mix up a bucket of paint and head out to touch up one of his old masterpieces. After he retired, he did it only when he felt up to the task, but it was still a labor of love.

"If I could've made a living elsewhere," he noted, "I would've painted these things for nothing."

WONDER why this Mail Pouch ad has weathered the years so well? Because it's new! The owner, a farmer in upstate New York, re-created one of the original Mail Pouch designs with help from family and friends, says Niles Eggleston of Milford, New York.

NANCY GARCIA of Superior, Montana had seen this barn many times when traveling on Highway 62 in southern Indiana but was inspired to photograph it when she noticed its striking new paint job. "We don't see many, if any, painted barns in Montana," Nancy points out. This beautifully maintained bank barn, which was built in 1904 and restored in 1993 by a group called Lanesville Heritage Weekend Inc., stands just east of Lanesville, Indiana.

MAIL POUCH BARN in rural Ohio caught the eye of Lia Jones of Enterprise, Alabama.

OUT of more than 70 photos that Doris Hunger has taken of Mail Pouch barns, this is the only one with the wording "Chewing serves to steady nerves". Looks like the painter could have used some! "This barn is on Route 286 between Glen Campbell and McGees Mills, Pennsylvania," writes Doris of Punxsutawney. "For the last 3 years, my husband, John, and I have been taking photos of every Mail Pouch barn we can find."

WHEN Shirley Conrad of Salem, Illinois spots a Mail Pouch barn, she always brakes and reaches for her camera—and so does her sister, Phyllis, who snapped this photo and shared it with Shirley.

CARPENTER Larry Sedlmeier of Woodville, Ohio altered the usual Mail Pouch ad to reflect his profession. "I did it just for fun; many folks don't even notice the difference," Larry writes. Mary Williams of Toledo photographed Larry's barn, which is located on Highway 20, between Toledo and Fremont.

BLACK BACKGROUND behind Mail Pouch signs is unusual, as is the fact there are two ads on this barn on U.S. Route 36 near Coshocton, Ohio, says Wendell Fink of Fredericktown.

"OH, my achin' back!" this barn near Saratoga, Iowa seems to be saying. Connie Knode, who lives about 2 miles away in Riceville, says the barn was built around 1950 with homemade rafters and has been sagging badly in recent years. "It probably won't last too much longer," she observes sadly.

PASSING this brightly painted dairy barn on her way to work each day keeps Kimberly Moss' spirits up. "It always reminds me to smile," explains Kimberly, who lives in Mukwonago, Wisconsin.

Barn Faces

Sometimes barns make people feel so good that the barns themselves can't help but smile. At least, that's what we're guessing after seeing these photos. Whether they're surprised, happy or even giving us a sly wink, it seems that even barns need to express themselves every once in a while!

EVERY HALLOWEEN, this barn on Highway 1275 near Monticello, Kentucky is transformed into a big "jack-o'-lantern", reports Carolyn Bertram of Monticello. At night, candles light up the barn's "eyes".

The "smiling barn" south of Nampa, Idaho was a landmark in our area for years. Every time I drove by, it seemed to be leaning a little more, making its "smile" a little wider.

After living in Idaho for 50 years, I moved to Wyoming. But on my annual trip back home to visit relatives and friends, I always drove by the old smiling barn and was amazed that it was still standing.

My last trip to Idaho was in July 1995, and the first thing I did was drive out to see the smiling barn. I couldn't believe it was still there! It was leaning more than ever, with an even wider smile, as if greeting everyone who drove past on that sunny day.

I felt such respect for that barn and finally took a photo that day. Time and time again, it seemed to speak to me, saying that most things in life can be conquered with a smile.

It was fortunate that I took the photo when I did, because 2 days later, a 65-mph windstorm hit the Nampa Valley and the barn finally collapsed. I will sorely miss that wonderful old landmark, a piece of Idaho's history.

—Anna Nordstrom
Sheridan, Wyoming

"SMILING BARN" on 12th Avenue Road, south of Nampa, Idaho, seemed to get happier as it aged, its smile widening every times its walls shifted. The barn was leveled by a windstorm in July 1995, just 2 days after Anna Nordstrom of Sheridan, Wyoming took this photo.

CHEERFUL BARN (right) near Savonburg, Kansas is always a welcome sight for Sheila Schlotterbeck's family. "I don't know who owns the barn, but we pass it every weekend on our way to Missouri," says Sheila, who lives in Chanute, Kansas.

COMEDIAN Oliver Hardy may have provided the inspiration for the face on this barn near Delaware, Ohio. LeRoy Schultz of Morgantown, West Virginia says the owners wanted to attract buyers for their homemade sausage, sold inside the barn.

STACK of hay bales appear to make this barn (left) in Idaho mighty happy, surmises Beverly Koetitz of Stanwood, Washington.

THE OWNERS of this barn (right) on Highway 135 between Kilgore and Overton, Texas enhanced its wide-mouthed "face" by painting on a set of eyebrows. Lorene Lewis of McLeod provided the photo.

BARN with lots of character appears to be surprised by its aging condition. Photo was snapped by Richard Gann of Eaton, Colorado on Highway 9 on the way to Clear Lake, Washington.

"HAPPY FACE" barn is located somewhere between California and Canada, reports Judie Lepisto of Santa Rosa, California. She says she's not sure about the location because she snapped the photo during a long trip—"17 states in 10 days"!

WEATHERED "face" was captured on film on a rainy spring day in 1995 by Steve and Jill Flinn of Keytesville, Missouri. The barn is near Forest Green.

"ABANDONED BARN is located approximately 12 miles west of Peetz, Colorado," writes Jim Waddle of Colorado Springs. "Maybe it was shocked that I was taking its picture!"

THIS barn in the Blue Ridge Mountains near Vilas, North Carolina has weathered everything from floods and ice storms to Hurricane Hugo. "It's one tough barn and has every right to smile," says Carlotta Tobia of Wilmington, North Carolina. "It's also one of the larger barns in the area." Mrs. Ramon Navarro of Tampa, Florida shared the photo.

AGAINST weathered gray siding, the white doors of this gambrel-roof barn near Winston-Salem, North Carolina look like a row of teeth. Leota Huff of Commerce City, Colorado shared the photo.

ONE-EYED BARN was spotted by Richard Adam of Monroe City, Missouri on Highway 1 in southern Illinois, near Gibsonia. Maybe the one eye explains why the barn seems to be scowling!

"I PHOTOGRAPHED this happy little barn near Atlantic, Virginia, near my brother-in-law's farm," writes Joanne Clifford of Orange, Connecticut.

MELBA EYLER has long been fascinated with "barn faces", and this is one of her favorites. Melba, who lives in Seattle, Washington, says it's been especially enjoyable to meet barns' owners over the years. "I wish you could have seen some of their expressions when I asked permission to take a photo of their barn because it had 'a good face'," she relates.

Chapter 3

Barn Yarns

HIGHEST WINDOW in this barn (shown here in the late 1800's) provided James Dufour with a great view of the surrounding countryside.

The most memorable part of my grandfather's farm was the big red barn, which looked you square in the eye as you came up his driveway. It sat majestically above all the other buildings, ruling the farm from its hill.

When I was a child, the barn was like a castle—a world unto itself. Exploring it was pure pleasure. The light, colors, sounds and scents are so imbedded in my memory that I only need to close my eyes and think about them, and suddenly I'm a child again.

The milking area was the most pleasant part of the barn on cool mornings. It was sweet-smelling and warm, with so many soothing sounds—the "maaaws" of hungry calves, the rustling and swishing of feeding animals and the muffled clinks of stanchion chains.

On a sunny day, when the double doors were open, the haymows and lofts were bright and golden, filled with cascades of billowing hay. Sparkling particles floated in the air like fairy dust.

If the doors were shut, the mows and lofts loomed over the barn floor, whispering of mystery and adventure. Streaks of sunlight penetrated the darkness through holes in the roof and cracks in the walls. Towering piles of hay became forts for us to play in.

We'd climb from the barn floor up through the massive haymows, then up an old wooden ladder to the center loft, just below the cathedral-style roof. We'd skirt the big square hayfork hole and climb over the hay to reach our "lookout"—the highest window in the barn.

This was the best place in the world to be on a sunny summer day. We could see the fields in all their shades of green. It seemed as though we could reach out and touch the stone walls bordering the farm, or the woods adjoining it. From there, we were one with the land.

I often fell asleep there in the warm sun, enveloped by the clean aroma of fresh hay. That was my idea of heaven. I remember sitting there, over 30 years ago, and thinking, "Someday, I have to write about this. This is too nice, too wonderful, to lose."

—*James Dufour, Penacook, New Hampshire*

When I was 11 years old, I built an airplane from sapling poles and wooden orange crates, and covered the wings and fuselage with linoleum my mother had discarded.

The plane had a steering wheel, borrowed from an old car, but no other controls. And although I had no idea what made a plane fly, our barn's long slanting roof seemed like a good place to take off.

Well, the plane did *not* lift off. Instead, it dove right into a new-blown straw stack in the barnyard, with me inside! At least it was a soft landing.

Dad always wondered how the shingles came off the roof above the haymow, but he never did find out. By the way, I finally did become a pilot—at age 64! —*James Baughman, Findlay, Ohio*

"MY FATHER'S BARN near Elida, Ohio is built to last; it has withstood many storms in its 65 years, even though it's one of the tallest barns around," writes Janell Stockton of Elida. "Mom and Dad take good care of the barn; they've painted it four times by themselves. That's Dad (Karl) with grandson Kyle."

FAMILY BARN built in 1912 provides a backdrop for Esther Van Vugt and her dad (on the second horse from the right) in 1931. "If the barn could talk, it would tell many interesting stories," she notes.

My grandfather homesteaded on the plains of North Dakota, near Strasburg, and built a barn there in 1912. Horses and cows occupied opposite ends of the barn, with a large hay area in between.

My parents moved to the farm when I was 6. We children made tunnels in the loose hay and climbed to the haymows to put on "shows", play house and make mud pies. In winter, Dad would tie up a rope so we could swing from one haymow to the other.

When the horses were being harnessed, we'd climb up on the stalls to watch, and sometimes we'd sit on the gentle ones. The harnesses always smelled of new oil. Even the whiff of ammonia from the manure smelled good to us farm kids.

At milking time, we'd watch our parents sit on T-shaped stools, aiming the foamy milk into pails balanced between their knees. It looked so easy—until *we* had to do it!

Milking was always a good time to sing, whistle and listen to the cows' contented chewing. The kerosene lantern would cast dancing shadows as the cats played in the straw bedding, waiting for a squirt of milk. In summer, we'd watch the returning birds build their nests in the rafters.

After chores, we'd often climb to the barn roof, where we could see for miles over the open plains. We'd sit there in the evening shade, singing and calling out "hello-o-o-o", then waiting for our words to echo back to us. It was like having a secret friend over the hill somewhere.

"THE BARN on my father's ranch near Bassett, Nebraska was built during the early 1900's," says Mary Jo Davis of Ainsworth. "I remember playing in the haymow as a child, milking cows and searching for new kittens."

As the years went by, the farm stood empty for a time. After I got married, my husband and I purchased the farm from my parents and restored the barn. While our three children were growing up, they enjoyed it just as much as I had, riding their ponies and playing rodeo while we milked.

In 1963, we left the farm for the mission fields of Nigeria. We had a large auction and rented the farm out. Now the barn is empty except for birds and pigeons. There are no sounds but the creaking of doors and the wind blowing through broken windows.

But if I close my eyes, I can again hear the voices of children at play, the cows mooing gently and the horses munching their oats. And if I'm *very* quiet, I can even hear an echoing "hello-o-o-o".

—*Esther Van Vugt*
Grand Rapids, Michigan

CLASSIC dairy barn with a matching outbuilding near Lake Benton, Minnesota was captured by Judie Lepisto of Santa Rosa, California.

Our family's treasured old barn was built in the 1920's on my great-grandfather's farm in Dickinson County, Kansas. Our mother was 4 or 5 at the time and remembers her grandfather and the neighbors doing the carpentry.

The barn stood for many years, housing horses, cattle, sheep and even hogs under the same roof. Eventually, poles were propped against the sides for some time. When even they couldn't support the walls any longer, we knew it was time to tear the barn down.

One of my brothers-in-law made picture frames from the barn's wood, and I enlarged some old photos of the barn and our horses and cattle. For Christmas, all six of us children received a set of framed photos, as did our mother. Now each of us has a remembrance of the barn in our own home.

Our own children enjoyed and played in the barn, too. One of my nieces, Jennifer Lexow, wrote about it for her high school English class. Her story (which is printed below) speaks for all the generations of our family who knew this old barn so well. —*Linda Melton Stockton, Kansas*

It wasn't a pretty barn, but I loved it just the same. The boards were rough and beaten from the constant harassment of the years, and the hayloft ladder rungs bore indentations left by millions of trips.

The barn leaned so badly it wasn't even safe to go inside. Every time there was a windstorm, we wondered if it would still be standing in the morning.

My mom grew up in that barn. I've heard numerous stories about the times she spent there with her brothers and sisters. I can just imagine the wild times they had.

My cousin Phillip and I came along a few years too late. The hayloft was too dangerous to play in, and later the barn was off limits completely.

That didn't stop us, though. We spent a lot of happy hours in that barn. Whether we were discovering furry little kittens tucked away in the corners, or exploring every nook and cranny, we always found something to do.

Sometimes when I was alone, I liked to stand in the passageway. As the sunlight peered through the cracks, lighting up the dusty crevices of the old building, I envisioned another time—a time when my grandfather was still alive. I could almost see him busily going about the morning chores, whistling a tune to himself.

Last Mother's Day, our whole family gathered to tear down the old barn. In the end, I felt a great emptiness, like something was gone that could never be replaced.

I thought of all the barn had seen. It had been through the good times and the bad. It had witnessed the Depression and the coming of automobiles, telephones and everything else our world is today. Births and deaths had taken place beneath its vast roof. The barn had followed the footsteps of many children and watched as they grew into adulthood.

As I watched the last walls of the barn crash to the ground, I was filled with wonder. Tears came to my eyes as I saw my grandma watching sadly; I wondered if she felt what I did.

It was then that I discovered that things will keep on changing and time will keep on marching along. Nothing can remain forever.

RUSTIC keepsake picture frames were fashioned from the boards of Linda Melton's family barn after it was torn down. The frame holds a photo of the barn, which used to stand in Dickinson County, Kansas.

During high school, I hated working on the farm. I joined every sport and club possible to keep me away from home. See, on the farm, there was always something that needed to be done, and Dad never let us take a vacation. "The livestock don't take a vacation," he'd say. "They still need to eat."

But a funny thing happened when I left the farm to go to college in the city. I wanted so badly to go home! I wanted to help Dad bale hay and climb to the hayloft and feed the cattle—even feed the hogs. It wasn't until I left that I realized what a heritage I had.

When I came home at Christmas, the barn was the first place I went. To me, it wasn't just a place to house livestock, hay and machinery—it was a living structure. My eyes filled with tears as I thought of all the good times I'd shared with this old friend.

I remembered putting up hay, my favorite chore. I'd drive the tractor for the baler, then put the bales on the elevator to the mow. After a wagonload was put up, we'd sit for a few minutes in the breeze that blew through the open doors on the ends of the barn.

I remembered feeding livestock with Dad during the blizzard of 1978. We couldn't see a thing—except for that great barn, a haven from the bitter cold. Even when the wind howled through the siding, I wasn't afraid. I knew my friend would protect me.

I remembered building so many forts and tunnels in the hayloft that Dad couldn't find me. I spent hours there, reading Hardy Boys mysteries and telling the barn my secrets, just as my mother and her siblings had done before me.

In the summer of 1988, my father passed away, and the hayloft was my refuge. Most of my best memories of Dad had taken place in the barn, or revolved around it.

About 2 years later, Mom sold the farm, and again I sat in the barn and cried. The farm had been my life growing up, even though I hadn't wanted to admit it then. The new owners invited me to stop in whenever I was home, and I did. The barn was always the first place I went.

Now the barn is gone. The owners needed a new dairy barn, and the old one was torn down. I have a few mementos—the weather vane, some slate from the roof, a beam, a piece of siding and a wooden peg. Those things may not sound like much, but they represent childhood memories of my friend, the barn, that no one can take away.
—*Linda McElroy, Springfield, Missouri*

The foundation of our barn was partly exposed on the south side. During winter, I would sit on that foundation in the warmth of the sun and find true serenity.

I used to stay there for hours, watching the cows in the corral or just dreaming as I watched fluffy clouds float by. Whenever I need to find that peaceful feeling, I just close my eyes and sit again in that warm winter sun.
—*Jeanette Urbom, Overland Park, Kansas*

"WE'RE PROUD of our old barn, which began as a log cabin built by John P.M. Higgans in 1856," reports Arthur Burger of Forest Lake, Minnesota. "The first white female in this territory was born in the cabin in 1859. The cabin was expanded to hold livestock and later became a Pony Express stop. The roof is made of the original hand-cut cedar shingles, and the logs are still held together by wooden pegs."

THIS BARN bore owner William Maxwell's name until 1975, when his grandson, B.C. Maxwell, became the owner. B.C.'s daughter and a friend painted his name there. "They somehow climbed up inside the barn and out through the round window above the lettering," recalls B.C.'s cousin, Hugh Maxwell. "It's a wonder they didn't break their necks!" Family reunion photo (lower left) was taken shortly afterward.

O ur grandfather's barn, built around 1890, had a traveling dolly at the peak of the roof. It had ropes and pulleys that extended to the floor and were used to lift hay to the loft in a sling. With a normal wagonload, you could lift one-third of the hay at a time.

One day, my dad and his brothers were showing off for their oldest brother and tried to lift *two-thirds* of the load. As they soon found out, this was a big mistake—the roof started to collapse!

The incident left a permanent sag in the roof (see photo at top right). Despite this, however, the barn remained in use for another 40-plus years.   —*Hugh Maxwell, Clare, Michigan*

S hortly after the turn of the century, my grandfather, William Swartz, decided he wanted a round barn on his farm in Fayette County, Pennsylvania. He'd never seen one, so he drew up his own plan, built a scale model and used it as a construction guide.

When he finished in 1905, he had used 5,500 feet of framing lumber, 3,800 feet of house siding, 40,000 cedar shingles, 2,800 feet of 1-inch oak boards, three kegs of spikes, six kegs of nails and eight pairs of hinges. The barn stood 60 feet in diameter, and cost $1,568 to build. The barn was beautiful—people from all over the country came to photograph it. But it had one drawback. If *just one cow* decided not to go into its stall, there was no way to head off the rest of the herd without help!

This barn also figured in the memorable day my father decided to learn to drive. He climbed into our 1920 Liberty touring car with my uncle and a friend while I hopped in the backseat.

Dad did fine—until he had to steer the car between the barn and another building. Instead of turning the wheel, he stepped on the gas, sending the car careening straight toward the barn. Dad, who was more accustomed to driving horses, pulled back on the steering wheel and yelled, "Whoa, whoa, whoa!"

Of course, that didn't work. The car just missed the barn door before slamming into the wall and knocking a few stones out of the foundation. A duck nesting in the tack room on the other side flew out the door, squawking her head off. The falling stones had crushed her nest, and she looked as if someone had dumped a bowl of beaten eggs over her head!

We had a big laugh about the duck, but my father wasn't too happy about the damaged wall. He never did learn how to drive after that.   —*Mildred Osborne, Uniontown, Pennsylvania*

"MY FATHER wasn't blessed with any sons, so my sisters and I did our best to help out while growing up on our 80-acre farm in Wisconsin," says Anita Brown of Brookfield. "We had a unique way of painting our barn. The tallest of us would paint as high as she could reach, then the next tallest would finish painting that board as high as she could reach and so on. By the time we were through, everything except the upper 3 feet was painted; Dad would finish the rest." (That's Anita at age 6 on the far left.)

One day in the early 1950's, I watched from our farmhouse window as the tail of a tornado hit our barn. It tore off part of the roof and sent the two big barn doors whirling to the ground in splinters.

After assessing the damage, my father called the insurance company. Within a few weeks, the company sent us two new doors about 12 feet high and 6 feet wide. "Tomorrow, before you and your buddy Ronnie go fishing, I want you to cut the new doors to size and hang them," my father told me.

The next morning, Ronnie and I got out the extension ladder, set it against the barn and proceeded to measure the height from the ground to the roller rail at the top of the door opening. Then we cut 2 feet off the bottom of each door.

As Ronnie lifted the modified doors so I could place the rollers on the rail, I yelled, "Lift the door higher—I don't have the rollers on the rail yet."

Ronnie replied, "What do you mean, lift it higher? The bottom is already 2 feet off the ground!"

The doors would have been a perfect fit—if we hadn't cut them. We hung them anyway, with a big gap at the bottom.

The wait for my father to come home from work that day was

"WHAT'S SO SPECIAL about this barn?" asks Connie Moore of Medway, Ohio. "It's stained a deep gold hue that shines in the sun. It sits on raised ground so on sunny days, it can be seen from a long way off. It's beautiful."

BARN DOORS that were cut too short still hang proudly on the farm where Earl Hensel grew up near Bath, Ohio—a reminder of a mistake that had his dad fuming!

the longest of my entire life. When he saw our handiwork, he sputtered a series of words I'd never heard before.

But the doors stayed. In fact, when the farm was sold several years ago, the doors (above) were still in place, just as we'd hung them all those years ago.   —Earl Hensel, Franklin, Tennessee

When I was 3 years old, Dad came up from the barn one night to get me so I could watch him milk. Every detail is imprinted on my memory.

The air in the barn was heavy with the aroma of hay and animals. Dad got a T-shaped milking stool for each of us, and I struggled to balance on the little two-board seat. The bare bulb overhead provided an oasis of yellow light amid the barn's vast darkness. A procession of our pet farm cats strolled into the light and arranged themselves in front of me, forming a half-circle around my father and the cow. Dad pretended not to notice.

As the first squirt of milk splashed into the bucket, the cats sat up on their haunches, front paws in the air. Without missing a beat, my father aimed a stream of milk at each cat's mouth. One after the other, they all got a drink, then sat down to lick their sticky white whiskers and faces.

I sat and watched, enraptured—and amazed at my father's skill. Dad gave me a big warm smile over his shoulder, then turned his attention to the milking. I didn't say a word. I didn't want to scare the cats away.

The following year, my dad died and I moved to town. I returned to the barn only once, more than 20 years later.

On the day I went back, the summer breeze carried the dreamy scent of sweet clover. I walked into the barn, to the corner where Dad and I had gone to milk. A haze of yellow sunlight came through the top half of the Dutch door, lighting the space.

To my surprise, there sat the two milking stools, just where they'd always been, propped against the wall. I closed my eyes. I didn't say a word. Once again, I was with my dad, the cats and the cow...in our old barn.   —Judy Nissen Boelts, Fairbanks, Alaska

The old red barn had already been demolished by the time I got there. It never would've survived the move with the farmhouse anyway. I stood there, recalling the lessons the barn had taught an awkward preteen many years before.

I recalled the magical moment I discovered a nest of wriggling

YOU CAN'T MISS the Shepherd's Farm Barn along Route 281 near Confluence, Pennsylvania, says John Whalen of Connellsville, who took the photo. Built in 1904, the landmark is 55 feet high, says owner Shirley Knoblach. "The farm got its name because we raise sheep," she notes.

"IN 1902, my grandfather built this barn as a wedding present for my parents," writes Marlene Schroeder, Newton, Kansas. "They lived in it for 8 years, and their first four children were born there."

newborn mice. I recalled the satisfaction of finally getting my feet and legs through the gymnastic rings above the hayloft and swinging gently, with no sounds but the mournful melody of doves and the uneven stomping and chomping of our old horse.

And I remembered the day when, riding bareback, I learned there's no stopping even an *old* horse when he picks up speed and heads for his own barn. Now I, too, know what it means to hurry home and take pleasure in familiar surroundings.

Reverence for life, perseverance, the need for quick reflexes, an appreciation of home—I learned about all these things and more in and around the old red barn.

As I was leaving, I noticed one old gymnastic ring, barely visible in the dirt where the barn had stood. Somehow it seemed appropriate that it be buried there. —*Ruth Gunther, Valparaiso, Indiana*

In February 1913, my grandparents bought a farm near Erskine, Minnesota that had a barn (right), but no house. They moved into the granary with their four children until spring, when they started building a house.

By the time my par-

DURING thunderstorms, Sharon Rudolph's family used to flee from their shaky old house to the relative safety of this old barn.

ents returned to the farm with four children of their own in 1941, the house was in bad shape. During thunderstorms, we fled to the barn because we feared the house might blow down!

Mom would awaken us and throw coats or blankets around our shoulders. Then we'd run to the barn in bare feet and nightclothes and wait out the storm. When it was over, we'd traipse back to the house through the mud. Mom would wipe our feet and we'd jump back into bed. I don't remember ever being scared; it was all rather exciting.

Around 1947, we got a "new" barn—one we bought from a neighbor and moved to our place with tractors and trailers. We continued using the old barn for sheep—and, of course, as our "storm shelter"! —*Sharon Rudolph, Annandale, Minnesota*

My great-grandfather built our barn in the 1860's, and it was huge. It included a wagon shed, granary, harness room, three haylofts, horse tie-ups, cribs and cow stalls—*plus* a henhouse and a pigsty!

The beams were hand-hewn with an adz and held together with wooden pegs. I used to climb to the highest beam and do double—sometimes triple—somersaults into the hay below. What a thrill!

I loved listening to the twittering of the barn swallows, who made their nests high in the rafters, away from the barn cats.

One of my favorite barn chores was feeding lambs the ewes wouldn't nurse. Grandmother would make them bottles of warm milk laced with a little coffee. The lambs would butt the bottle and swivel their tails in delight! —*Priscilla Osgood, Bangor, Maine*

61

 VI JANSICK'S two older sisters and a brother were born in this barn, shown in the 1930's (inset) and in 1995. "Note the tree growing out of the silo!" she points out.

W hen my nephew was small, his Sunday school teacher told the Christmas story, pointing out that the Baby Jesus was born in a stable. "I'll bet none of *you* know anyone who was born in a barn," the teacher said.

"I do," my nephew exclaimed. "My mom was!"

Imagine the teacher's surprise when she repeated the story to my nephew's mother—who was also the pastor's wife—and learned that it was true.

Our parents had very little money when they started farming near Bingham Lake, Minnesota in the early 1930's, so they built their barn first and lived in half of the divided haymow. It was several years before they had the time or resources to build a house, so *three* of my siblings were born in the barn.

Whenever somebody jokingly asks them if they were "born in a barn", the answer is always a resounding "Yes, as a matter of fact, we were!"
—*Vi Jansick, Fridley, Minnesota*

I n August 1940, my parents sold their Colorado homestead and moved all seven of us children to Idaho. For 3 months, we stayed with some generous cousins and an uncle—16 people under one small roof!

My parents soon bought some land nearby, but the only building was an old loafing shed that would house the handful of recently purchased cows. Since having our own place to live was a priority, they decided to construct a simple barn and live in that un-

til we built a house. So, for the next 2 years, the barn was our home. It wasn't large, and there was no electricity, so our nights were lit by a single kerosene lamp. Its light, along with the heat from a wood cookstove, seemed to instantly drift off to the open rafters above us.

The walls, a single layer of rough-cut boards, offered little protection from winter's cold. Wind and snow blew right through the cracks. We stayed as close to the stove as we could; if we stepped even a few feet away, we'd quickly feel the chill.

After 2 years, we moved into a chicken house Daddy had built. In summer it was unbearably hot, since the cookstove was always being used for meals, laundry, baths and ironing.

We planned to live there only temporarily, while Daddy and Uncle Glen built our house. But with World War II going on, building materials were in short supply. And since my parents didn't believe in borrowing money unnecessarily, the house went up a few boards and nails at a time, as finances permitted.

When finished, the house was extremely modest by today's standards—and probably those of 1948. But to us, after living 2 years in a barn and 6 years in a chicken house, it was a mansion!
—*Margaret Crill, Nampa, Idaho*

"OUR BARN was built around 1900 by a man named William Silverthorn," reports Jean Smith of Yale, Michigan. "He used beams hand-hewn from trees that grew across the street from the barn site. My mom and dad purchased the farm in 1943, and my husband, William, and I bought it from them when they retired in 1973. It's one of the oldest and largest barns in the area and considered a local landmark."

W hen I was 11 or 12, a bunch of kids were in our hayloft, dropping hay out the window to our Jersey bull, "Buck". Then someone got a bright idea. What if we took the rope attached to the hay tongs and lassoed the bull's horns, just like in the movies?

We all took several turns, with no success, dropping hay all the while to keep Buck in the same spot. When it was my turn again, I put a large loop in the rope, and the noose fell over Buck's neck! He didn't like that much, and started backing up until the rope was tight around his neck.

There was trouble on our end of the rope, too. The rope was connected to a track mounted in the rafters, and the barn was beginning to crack and pop!

Dad was working just 100 yards away, and when I ran to tell him what had happened, he was madder than the bull. He raced to the barn and tied a snapper to another rope. Then he slowly walked over to Buck and hooked the snapper onto the brass ring in his nose.

With two fingers curled through the ring, Dad led Buck back to the barn and cut the lasso. As long as Dad was holding the ring, Buck was under his complete control.

Dad led Buck to the bullpen and shut the door, leaving just enough room to reach in and remove the snapper. The instant he let go of the ring, all Buck's fury surfaced. He ran, bellowed, pawed, snorted, rolled and ran some more. Dad laughed until he cried, but we still caught it for that stunt. That was my last attempt at a rodeo. —*Al Cothran, Summerton, South Carolina*

S hortly after hayforks became available, my father installed one in the west gable of our new barn. The fork could pick up a quarter-load of hay with one thrust, then it'd travel along a metal track under the rooftree.

There was something else new in the barn, too—dynamite. The men used it to clear stumps at the edge of the woods, and the unused portions were stored in the barn's southwest corner.

One day a horrific electrical storm came up unexpectedly, and the men hurried home, their horses in full gallop. They had just gotten the team into the barn when lightning struck the hayfork.

The horses panicked, the men were knocked to the floor and the barn filled with smoke. The electrical charge streaked down the hayfork's metal track and shot through the roof, searing a path through the wooden shingles. Then it plunged to the ground, cracking the foundation.

Thankfully, nothing ignited. I think that's because my mother was in the house the whole time, frantically praying!

Today, the barn still stands as sturdily as ever after 80 years of use. My father built it to last, and it did.

There's an unusual story about the hay shed, too, although I'll

AL COTHRAN got into some mischief in the barn shown here behind his late father-in-law, Richard Crouch.

never know why it was called a "shed". It's huge! A team with a load of hay could drive through the center.

What's unusual about this building is that the ridgepole is supported by living trees denuded of bark and limbs. Over 100 years later, the roof is still straight as an arrow.
—*Lillian Zajicek*
Belleville, Illinois

M y father had owned our southern-Indiana farm for only 4 years when lightning struck his barn in 1926. It was nighttime, so all 30 of our Jersey cows were inside, plus two teams of horses, all our farming equipment and a brand-new 1926 Overland touring car. Dad lost them all.

Dad kept farming and built a new barn in the same location. But in 1932, that barn was struck by lightning, too! I was there when it happened and will never forget it.

I was playing in the haymow with my brother, sister and cousin when Mother suddenly came in and said, "You children come to the house *now*. There's a bad storm approaching."

Just as we stepped onto the front porch, lightning struck the barn—in seconds, it was engulfed in flames. The hay door and hay track were thrown *400 feet*, and glass windows melted into puddles on the concrete floor.

Dad and a hired man raced toward the barn, hoping to save a hay wagon parked just inside the door. But when they pushed the door open, intense heat and flames roared out. There was nothing they could do.

It was a catastrophe—the second in 6 years—but Dad wasn't one to give up, even during the Depression. He drew up plans for a new dairy barn and his friends provided the lumber, nails, hardware and labor. In less than 2 months, he had a new barn with stanchions for 32 cows and we continued to farm and run our dairy.
—*Loren Stephenson, Osage Beach, Missouri*

My neighbor intended to burn down an old barn on a farm he owned—until a fellow offered him $1,000 for the weathered boards. The buyer said he'd remove the boards himself, then sell them to interior decorators for use in homes and restaurants. The deal was closed.

Meanwhile, my neighbor rented a farmhouse on the same farm to a young couple for a very low rate. They were so happy that they wanted to do something special for him. So they surprised him—by painting the old barn red!

The buyer no longer wanted the boards, the neighbor was out $1,000 and the young couple were out their paint and labor—because then my neighbor *did* burn the barn down!

—Herman Hoffman, Michigan Center, Michigan

I dreaded rainy days—that's when I had to ride the bus to school instead of walking. Our dilapidated barn was always the brunt of the other children's jokes.

"Hey, does your barn leak?" one would ask. Another would reply, "No, it pours!"

The barn leaned precariously for years. And when it finally succumbed, it fell so slowly the cows hardly seemed to notice!

When Daddy built a new barn, it was all I could have dreamed of—solid and compact, with four stalls and a roomy divided crib. The most fascinating part was the stairs leading to the loft, where I immediately established squatter's rights.

In the loft, I could spend hours with my dolls, "keeping house" to my heart's desire. The aromas of cotton seed, dusty corn, mice, mules and cows mingled in a pleasant, unique odor.

The new barn was also the perfect site for advertising signs. My favorite was Clabber Girl baking powder. Each time the company

AS A CHILD, Frances Cleary-Wittmeier dreaded her classmates' jokes about the family's rundown barn, shown behind her in this 1941 photo. When it collapsed, the new barn that replaced it became Frances' favorite place to play.

nailed a new sign to the barn, they gave Mama a cookbook.

—Frances Cleary-Wittmeier, Orange Beach, Alabama

I've known my neighbor's barn for 68 years, and it still stands straight and tall, none the worse for the wear.

I remember hot summer days when we'd go out to rake hay, then stamp it down on a wagon until the pile got wider and higher. We'd ride back home on top of the wagon and step right through the open doors on the barn's upper level. Pitching from wagon to barn would commence. We'd stay inside, stacking.

It was hard work, with no pay except the reward of helping our neighbor. When our work was done, we received a warm glass of milk straight from a cow—a real treat.

We'd also help harvest corn from the fields, shuck it and put it in a corncrib. Later, we'd put the ears in a large funnel-like machine and turn a wheel to remove the kernels. It was child's play for us, but the farmer, who also worked full-time in a foundry, was always grateful to find half a bushel of corn ready for feed.

We'd often sleep on the hay in that barn. No pajama party could compare with *these* parties! We'd eat candy and apples, tell stories and wear out flashlight batteries checking on barn mice.

On cold rainy days, we'd climb into the loft and snuggle deep in the hay, listening to the patter of rain on the roof. What comfort and warmth! Words can't describe the experience.

How I long once more to see the light in the barn at evening milking time, with a midwinter snow piled up outside. The farmer has been gone a long time now. But the barn keeps vigil, looking down on the old farmhouse like a faithful caretaker.

Spring's rain, summer's sun, autumn's melancholy and winter's ice and snow seem to have no effect on the barn. It's a landmark that will witness my own passing someday. It was a friend to me in my childhood, and I'm pleased that we're still neighbors after all these years.   —Geneva Bellinger, East Haven, Connecticut

The barn on my parents' 80-acre Ohio farm was my "special place", but to my parents, it was much more. The barn was the storehouse that secured our animals and our future. It was the lifeblood of our daily living.

Along with our cows, calves, sheep and horses, the barn held food, grain and herbs. Cabbage, potatoes, celery, carrots, apples and tomatoes that were stored in huge mounds of dirt stayed fresh far into February. For medical needs, Mother's special cupboard held dried wild herbs, leaves and tree bark, along with formulas for their use.

The granary held enough to feed our family and up to 20 ani-

THE SYMMETRY of these barns in Clare County, Michigan never fails to catch the eye of Kay Johnson of Harrison. "They capture my attention every time I pass them on my way to work," she notes. "I photograph them often."

"THIS BARN was a great point of interest for people in Bridger Valley," explains Glenys Birch of Lyman, Wyoming. "The barn, which is owned by the Town of Lyman, was built in 1933 and hosted many dances—sometimes with up to 400 people attending. The south end provides a majestic view of the Uinta Mountains."

mals for an entire year. The grains provided cereal for breakfast, at least 20 loaves of bread a week, assorted dinner and breakfast rolls and *144 biscuits* every morning. Mother even added browned ground wheat to the coffee to make it last longer.

In winter, the animals made the barn a haven of warmth, and we'd quickly shed our coats, scarves and gloves while doing chores. But before we would don them again for the walk back to the house, we'd have to search them for spiders—and *snakes*!

During the day, the main barn floor was like a community recreation hall. Big brothers, cousins and friends cleared the floor for basketball or wrestling. If we younger ones didn't get underfoot, we were rewarded with swift (and dangerous) rides on the hayfork.

The hard-to-reach space between the granary ceiling and the sloping tin roof was perfect for a private clubhouse or playhouse. My brother and I have often wondered about the secret space between two joists in the haymow. Does it still hold our treasure—a cache of dimes in a tobacco tin? Have other children added to our loot?

On hot summer nights, I used to take the dog to the side porch and curl up in my blanket. In the shadow of the barn, laying between the dog and a blanket, I felt as safe as if I was in Dad's arms.

The barn cared for all of us, animal and human alike, keeping things warm and dry. It still stands, grand despite its weather-beaten exterior. —*Morna Scher, Vero Beach, Florida*

Ever since I was a young girl, I've loved big barns and admired the men who built them. They created these majestic buildings—each one an architectural wonder—long before the days of high-tech engineering.

Each time I step into a barn to help with haying, buy feed, look at livestock or attend an auction, I'm in awe. Sometimes the families explain how the barns were built. Each one is unique, designed for its owner's specific needs.

I once was told that barns were built where they could be seen from the farmhouse kitchen; that way, the farmer's wife could keep an eye on things while cooking and canning. No wonder—barns were the key to the family's survival, holding thousands of bales of hay and straw, plus equipment and livestock.

Each time I see a barn in disrepair, my heart sinks a little. I know it isn't always feasible to repair them. They're becoming obsolete; even barns still in good condition have been bulldozed. And to think that not so long ago, the loss of a barn meant hard times for a family. —*Tara Miller*
Edinboro, Pennsylvania

The barn was the cathedral of Grandpa's farm, overshadowing every other building in both size and character—almost as if it had a soul. Grandpa kept it in excellent repair; Grandma even polished the windows.

Everything inside was built of heavy hardwoods, from the overhead beams and support poles to the mangers and planking on the stall dividers. With time, the wood grew scarred and dark, its sharp edges worn smooth by the animals. It was a welcoming, comfortable place.

Steep wooden stairs led to the haymow, with its floor of varnished 4-inch oak planks. The ceiling was so high our voices echoed if we hollered.

There were few windows, but light filtered down through the cupola in dusty rays. Like a church steeple, its very height caught your eye, inspiring lofty thoughts.

Though peaceful, the barn was never quite still. You could always hear the soft creaks and moans of shifting wood or the wind fingering its way through doors and windows.

I especially remember fall evenings when the frisky milk cows would come in from the pasture. They'd march to their stalls and Grandpa would slip a loose chain around each one's neck and greet her by name. As he'd settle in to milk, he'd lift his wire-rimmed glasses and rest his forehead against the cow's thickening coat.

Today, the countryside is dotted with run-down and caved-in barns—poignant sights for those of us with deep connections to a family barn.

—Joni Woelfel, Seaforth, Minnesota

Like all the farmers in Lawrence County, Tennessee, my father struggled to feed his family and livestock during the Depression. He was worried about the farm buildings falling into disrepair and was especially troubled by the rusting roofs on the barn and corncrib.

But God has a way of renewing all things! One day, while I was at school, a tornado swept through our farm, tearing off both those rusty roofs and part of the siding on our house. The wind-damaged barn was left leaning at a dangerous angle. We couldn't possibly house our animals there.

I'll never forget the

Bobbye Kudzma beamed with pride when she posed in front of the barn and corncrib on her family's Tennessee farm. Both buildings had been damaged by a tornado but were repaired by friends and neighbors.

look on Dad's face when I returned home that day. Our farm was the only one in the community that sustained any damage.

Suddenly other farmers began appearing, one by one, to survey the damage and offer moral support for "Uncle Jim", as everyone affectionately called my father. Over the next few days, these dear friends brought supplies, tractors, tools, horses and willing hands to reconstruct the barn, replace both roofs and fix the siding on the house.

The atmosphere was festive. I still remember the joy it brought to my teenage heart. Our neighbors weren't just renewing buildings—they were renewing friendships.

When all the work was done, I just had to have my picture taken in front of the improved farm buildings. I was so proud that my face almost outshone the reflection of the new roofs. Barns are not just barns. They're community builders!

—Bobbye Kudzma, Gaithersburg, Maryland

As a child, I loved looking at magazine pictures of pretty red barns. They didn't look anything like the ones I was used to seeing around our place in Palestine, Texas.

See, the barns around us weren't even painted. Most looked just like ours, with the lumber faded to a weather-beaten gray. My great-grandfather had built our barn before I was even born.

The barn may have looked ordinary, but it held many memories. My best friend and I loved to play there, climbing the stacked hay bales and pretending we were going "upstairs" to prepare for a ball. We'd descend the stairs gracefully, with towels pinned around our heads to represent long flowing hair.

I can still smell the yellow cottonseed meal my grandfather fed the cows. The meal cost more than cottonseed hulls, so he'd put the hulls in the trough first, then let me sprinkle meal over the top. Helping my grandfather made me feel so important!

—Verna Ray Humphrey, Palestine, Texas

Shortly after we were married, my husband and I returned home to central Indiana for a visit and attended a church get-together. As the evening ended, we noticed a group of our friends in a huddle, talking and laughing. We figured they were planning to shivaree us that night at my parents' place.

A shivaree was a noisy surprise attack outside the newlyweds' home after they'd gone to bed. We'd been married a few months and hadn't had a shivaree yet; we thought our time had come.

So we went back to my parents' house, gathered blankets, pillows and a flashlight and, instead of going to bed, headed for the barn loft to wait for the attack. Our friends would be in for a real

WELL-KEPT BARN with eye-catching paint job and ornate cupola catches the attention of passersby on Highway 53 near Eau Claire, Wisconsin, notes Marion Johnson of Greenville.

While visiting a pioneer village a few years ago, I decided to explore an old barn sitting off by itself. As I walked inside, the aroma of hay and horses brought back a flood of memories.

I was transported back to my uncle's horse barn in the 1930's, reliving an experience that was very precious to me. After the horses were unhitched from the hay wagon, I had been allowed to drive the team to the barn—a real treat for a 12-year-old girl.

My uncle removed the harnesses while I pumped water into a pail for the horses to drink. I recalled familiar sounds—the contented clucking of chickens in the yard, horses stomping off flies while slurping water, the harnesses jangling as they were hung on the wall and the rustle of straw underfoot.

Someone called my name; my impulse was to run to the house for a basket to gather eggs. Instead, I reluctantly returned to the present. I breathed deeply and enjoyed the memories for a few more seconds before they faded away.
—*Marilyn Golly*
Eagar, Arizona

surprise! After several hours, *we* were the surprised ones. No one came to shivaree us, after all. In fact, we never did have a shivaree—and we've been married 49 years!

That night in the loft was filled with sounds—horses and cows moving on the ground floor, mice scurrying, birds flying to and from the rafters. There were other noises, too—some of them scary. We didn't sleep much, but it's a night we still laugh about.
—*Lucille Stamper, Danville, Indiana*

In the fall of 1926, our old barn in Ohio burned, ignited by sparks from a threshing machine. The following spring, Dad decided to look for a barn we could tear down and move. He found one 12 miles away that had been built in 1897.

A neighbor marked the beams and joists before the barn was dismantled. The slate roof was carefully removed and taken along, as was the sandstone foundation.

The move took several weeks, and I made many trips back and forth with Dad, a wagon and a team of horses. We had a few memorable experiences, too. One man loaded too much slate in his truck and blew out all the tires!

Finally we were ready for the barn-raising. A man came from some distance to help lay the sandstone walls. Lots of neighbors and friends pitched in, and my mother and the neighbor ladies fixed the meals. For a girl of 12, it was quite an experience. I still live on the farm where all this occurred.
—*Esther Stima*
Mansfield, Ohio

When cousins or friends visited our farm, we often ended up in the barn, riding the manure carrier! We turned it upside down, straddled it and moved ourselves along by pulling on the cable.

The carrier had a latch to keep it from turning over when it was being filled. But when we rode it upside down, it swung freely and could buck us off if we weren't careful.

Our city cousins didn't like this treacherous "bronco", so they'd turn it over, fasten the catch so it wouldn't upset and ride inside. We farm kids knew better. Who wanted to sit on dry—or, worse yet, wet—manure? We'd rather take our chances and get bucked off!
—*Florence Foley, Leipsic, Ohio*

A MANURE CARRIER turned upside down inside this family barn used to provide Florence Foley and her daring friends hours of fun.

T he barn on the farm where I grew up is not uncommon. It's not high or wide or long, or oddly colored, or designed to draw praise or notice. It doesn't tower over vast acreage, and it doesn't seek to be more than what it is.

Logic would define it as an inanimate object, incapable of soul and spirit. But to me, it seems to be a living friend.

The barn draws its essence and character from the generations that have toiled there, just as the human soul is enriched by the barn's nature. This barn has known many generations, but I've witnessed only one—and the barn speaks volumes about the character of that man.

My father and the barn seem as one, inseparable. Together, they've weathered sun and storm with courage and grace, standing squarely on a solid foundation. Through the flow of life, they remain the constant—the hub of my family's wheel.

When our family was young, much was required of both man and barn. Through their efforts, the children learned life's lessons—honesty, a work ethic, respect for the balance of nature, self-worth and how to enjoy both hard work and recreation.

The children matured, grew self-sufficient and became parents themselves, using the lessons and values they had been taught. For man and barn, industry became secondary; now they could enjoy the security provided by a life of stewardship.

As another of life's seasons dawns, the roles will adjust again, allowing the rest and pleasure of retirement. I know that my father will succeed here as well…with a little help from his barn.

—Jeffrey Young, Appleton, Wisconsin

"THIS BARN stands on the Robert Rice ranch, about 5 miles east of Pendroy, Montana," reports friend LeRoy Schouviller of Huntley. "It was originally built by Dutch settlers between 1910 and 1915 and moved here by Robert's father, Paul. A painter persuaded Robert's son, Allen, to allow the white star on the barn; now it's lit during Christmas. The '2-K' is the ranch brand, which is still used."

G randpa's old barn stood massive against the June sky, a quarter of a football field in size. Weathered boards, sawed from oaks felled a century before, towered above us, held together with square metal nails forged by blacksmiths. The tar-papered roof pitched steeply against the sky.

With a heave backward, the heavy sliding door groaned open. We blinked at the sudden, almost spooky darkness and stood still, dwarfed in a doorway that once accommodated horse-drawn wagons stacked 15 feet high with hay. As our eyes adjusted to the dimness, we raced into the cavernous upper story, our feet echoing on the solid oak floor.

One by one, we climbed the handmade ladder to the haymow. The smooth rungs are narrower in the center, worn from decades of twice-daily trips for feed. We flopped face-first into the hay and came up laughing. Our shrieks replaced the ancient echoes of "gees" and "haws" and the friendly shouts of bygone harvesters.

We lay on our backs, staring at huge beams 20 feet above. Fitted together with hand-carved wooden pegs, they were wide enough for a man to walk across.

As we sat up, gazing at the opposite mow a basketball court-length away, we spied a door to a basement. We leaped off the mow wall and raced for the door, threading our way through old cultipackers, buck rakes and grain drills.

The door creaked open and we groped down the steep hand-built staircase into the cool darkness. We inspected Grandpa's old tractor, then wandered down a long aisle between wood-slatted pens. When Grandpa was young, the cows and pigs housed here warmed the barn with their heat.

At the front of the barn, we stepped over the high sill and outside, under the overshot. Nine wooden pillars support the overhanging upper section, and each has a 200-pound stone base hand-hewn somewhere on the farm's 200 acres.

Our minds were filled with Grandpa's stories of the old days…of Great-Grandpa shoeing huge workhorses in this very spot…of colts racing in the nearby pasture…of calves being born and neighbors helping bring in the harvest. They're sad stories, funny stories, heartwarming stories, all centered around this barn, once the focus of his family's life.

—Judith McKrell, Russellton, Pennsylvania

I hadn't seen our old barn in Tyler, Texas in a long time, and when I finally visited it again, I was surprised that it still looked the same. So many things look smaller or less important after a long absence—but not this barn.

Opening its oversized doors took me back into a world of yesterdays—like the cold February night when our first colt was born.

My mare had looked ready to deliver for a week, and we'd been setting alarm clocks to remind us to check on her. Finally, sons Trey and Michael telephoned me from the barn. The colt was coming!

By the time I ran the 100 yards to the barn, "Badger" had already arrived. Trey was sitting in the corner of the stall, his face full of pride and wonder—his first delivery! Michael wiped the colt dry as the mare inspected the new arrival, making soothing, reassuring sounds I'd never heard from her before.

It was cold in the barn, but none of us noticed. We watched spellbound as the wet colt tried to stand. He'd balance on three legs, only to have the fourth fold in the wrong direction. Sometimes, in frustration, he'd simply plop on his bottom like a puppy.

We laughed until tears rolled down our cheeks, and cheered and clapped when he finally made his first wobbly steps toward his mother. We left them just as the sky began to turn a soft pink—and only then realized our fingers and noses were numb.

I walked on, through the tack room that still smelled of leather, medicine and soap. Then I stepped into the long tractor room, where our sons had hosted so many gatherings—4-H get-togethers, Sunday school parties, Sheriff's Junior Posse Drill Team hayrides, campouts with friends. Those events were filled with joy and laughter, mingled with a few tears and disappointments.

As I closed the heavy doors and started up the hill to the house, I was sad to leave the barn and all those memories behind. But the barn's not going anywhere. I can always come back...and make new memories for tomorrow.

—GeNeil Avery
Tuscaloosa, Alabama

OLD CHURCH was bought by Rudolf Carmann and moved 3-1/2 miles to his farm, where he turned it into a barn. It doesn't look much like a church from the outside since sheds were added to the east, west and north sides, but inside, windows and a chandelier give away its origins!

When a Lutheran church near our Nebraska farm disbanded in 1955, my husband, Rudolf, bought the building for $600 and turned it into a dairy barn complete with stanchions, cement floors and a separator room.

After we added sheds to three sides of the building, it no longer looked like a church from the outside. But inside, you could still see the old church windows and a chandelier hanging over the stanchions.

Today, we use the building to store straw and hay. The roof is still good and the memories linger in our hearts.

—Dorothy Carmann, Riverdale, Nebraska

Dad gave dignity to our western-Oklahoma farm when he bought the old Stingy Ridge schoolhouse, moved it 2 miles north and turned it into a barn to store his wheat crop. All 11 children were proud of his skill in designing this barn.

On lazy afternoons, my sister Frances and I would climb the ladder to the hayloft to read storybooks, cushioned by the fragrant hay and lulled by the cooing of pigeons.

There was a stable on the north side of the barn with a deep wooden manger where we placed hay and oats for Dad's stallion, "Morgan". Long after Morgan was gone, the leathery scent of harness and horse collars lingered, stirring memories of his fiery temper.

Soon the Dust Bowl days were upon us, with their dark and discouraging times. But I'm thankful for the rich memories that preceded them.

—Marjory McCollough
Costa Mesa, California

SCHOOL became a barn on Flying V Stock Farm, says Marjory McCollough. No wonder she liked to read in the hayloft! That's her dad, Dow Jones, and his sons Louis (on horse) and Wilbur in photo.

THOUGH it's over 180 years old, this barn still is a productive member of Roger Wiles' farm, sheltering firewood and building materials and providing many memories.

For the past 6 decades, my family's old barn (above) has been part of my life. Now at least 180 years old, it's misshapen, creaky and broken. But it's still a part of my life.

When I was 2, I wandered away while Mom was watering our mare, "Flory". She knew Flory would return to her stall on her own, so she turned her loose while she looked for me.

Unbeknownst to her, I'd plopped down on the ledge leading to the stalls. Mom spotted me just as Flory stepped over me and into the barn—without touching a hair on my head.

When I was older, I played in the haymow with my brother and other kids, even though Dad didn't approve. I never understood why, until the day my brother chased me out of the mow with a pitchfork. For once, I didn't stay to argue with him, even though he was younger and smaller.

The barn housed our first tractor, a Farmall H, and later our Allis-Chalmers combine and tractor. It was home to goats, dogs, wild pigeons, chickens, Shetland ponies, cattle and, of course, barn cats. In recent times, opossums, skunks and raccoons found shelter among the few broken bales of straw that remain, now a quarter-century old.

Today the barn dutifully protects my firewood and building materials. I sometimes wish it still had the old grain bin, which was a great place to store squash and pumpkins; they wouldn't freeze when covered with grain. But I guess the barn has done enough for me already.

As long as someone finds it useful, I suspect the barn will continue its job—sheltering, protecting and providing memories.

—*Roger Wiles, Linden, Michigan*

"WHEN we bought this farm, our barn—built in 1915— was the biggest in our community; it could accommodate 50 cows," writes Freeman Davis of Wetmore, Kansas. "We've had people drive 50 miles to take pictures of our barn, and we're very proud of it."

"THIS BARN was built in 1913 by my father and grandfather," reports Russell Stewart of Monmouth, Illinois. "It was used to house Percheron draft horses until 1964. It's one of the largest barns in our area. Many people (including me) have had lots of fun swinging from the ropes in the hayloft!"

Of the many weathered barns along a rural road near our home, Grandma Mimi's huge gray one leans the most. Her father built it in the early 1900's.

On a recent drive, I stopped with my sons to look at "Mimi's barn" and shuffle through the ancient straw where she milked cows, shoveled what needed to be shoveled and dreamed her dreams.

The massive beams were still intact, holding up a cutwork roof. The barn had been vacant for half a century, but I could hear a pig-tailed little girl speaking to the mute cows, telling them of a life beyond the pasture.

My son found a horseshoe that had been tossed in a dusty corner. Had it been on the horses that pulled the sleigh for Christmas visits? Or was it from a horse that sidestepped Mimi's unshod feet?

The barn's enduring carpentry reflects only part of our legacy. My grandmother's character is stronger than those rafters. Hard work and financial deprivation prepared her to successfully raise three girls during the Great Depression.

As I see Mimi's barn become more fragile each year, I'm challenged to build a legacy in the hearts of my own children. They can repeat the stories of Mimi's childhood, my mother's and mine. Nurturing each other through hardships, delighting in our successes, we'll build a heritage as strong as the hand-hewed beams of Mimi's barn. —*Deborah Byrne, Chardon, Ohio*

During the Depression, we lived on the main highway northwest of Fort Collins, Colorado. Our old barn provided shelter to many people who were passing through and would offer to work in exchange for a meal and a place to sleep. I can't remember Dad ever turning anyone away.

Most of the travelers slept on gunnysacks in the large double-walled room that served as our granary after harvest. Dad provided plenty of old quilts for bedding.

I vividly remember some of the people who stayed in the barn. There were two boys just out of college who begged Dad for jobs and a place to stay; a family of seven who came through on their way to harvest potatoes in Idaho (my sister gave the youngest girl a doll); and a red-bearded gentleman who split firewood for us, even though he had only one arm.

I have often wondered what became of all the people who stayed in our barn. A few of them took the time to send us news about how they were doing, and I think a couple even sent Dad some money in appreciation of his help. —*M.J. Davidson*
Morro Bay, California

"MY FATHER, Joseph Rustine, still owns this barn in Monroe County, Pennsylvania," says Larraine Granacher of Stroudsburg. "It replaced one that was built by my grandfather but destroyed by lightning. My sisters and I spent many a rainy day playing in the haymow of this barn."

Excitement swelled in us as we dashed across the gravel driveway, through the backyard and around the garage. There it stood—the old rustic barn with faded brown walls, towering above the sheds Grandpa had once used so diligently.

We were awed at being so close to this magnificent monument. As we opened the 10-foot oak doors, the rusty hinges squeaked like those in a haunted house. The aroma of decayed hay engulfed us.

The cracked concrete floor hinted at days past, of feet whirling across it during barn dances. We glanced around, quiet and nervous, eager to discover the mysteries our castle concealed.

We scrambled to a 50-gallon wooden chicken feed drum that was suspended on its side between two lofty beams. Reaching into the top, we felt gritty seed mixed with dust that had settled over the years.

At last we came to the rugged stairs made of thick timbers. We climbed them slowly, one at a time, to the hayloft. In one corner, I recognized a splintered rocking chair. In the other corner was a tiny heap of straw. Bending over, I spied a cluster of tiny eggs nestled cozily in their mother's handiwork.

The nest was just one of the hidden treasures nature would preserve in this old barn. That awesome structure protected its tiny visitors even as it gave a third generation of children an adventure of the imagination. —*Charlene Mainz, San Antonio, Texas*

Chapter 4
Unusual Barns

"CHINESE BARN" near Deshler, Ohio was the subject of numerous photographs, drawings and aluminum etchings. This photo was provided by Erma Lee Crawford of Deshler.

Neighbors and travelers alike well remember the "Chinese Barn", which was located on West View Farm in northwestern Ohio.

George Hyslop designed this unique barn after visiting western cattle ranches to study their methods of fattening beef. It was built in 1910 at a cost of $11,500. That was an outrageous expense for that time, but Mr. Hyslop was determined to build the best barn in the land. It proved to be both an architectural and scientific marvel.

E.O. Fallis, a noted architect from Toledo, designed a roof with a central cupola that flowed into eight gables. He probably didn't intend to give the barn an Oriental look. More likely, the design was dictated by a ventilation system that required a five-story roof.

The barn measured 90 by 92 feet and was painted red with white trim. It had space for seven horses, 128 cattle, 40 hogs and 600 chickens, plus box stalls for foaling and calving. The floors were tongue-and-groove boards and the cedar roof shingles were held in place with copper nails.

There were about 182 rafters, and Mr. Hyslop hand-carved the outer extensions. On the advice of weather forecasters, Mr. Hyslop installed galvanized lightning rods, which required 500 feet of wire cable.

The design incorporated innovative methods for tending livestock—some devised by Mr. Hyslop himself—years before they were generally accepted. In an era of pastures, he fattened his livestock by confining them in comfortable quarters and bringing them feed.

At the center of the barn stood four silos 14 feet in diameter and 43 feet tall. They were filled by a central distributor that could be turned without stopping to reset the cutter. Silage was fed into a cart on an overhead track, then dropped through trapdoors to the pens below.

Spouts on the barn carried rainwater to a cistern, which fed water to the main tank. A float in the tank controlled the water level in every trough in the barn. This system failed only once, when ice formed on the water in the main tank, holding the float in place.

Mr. Hyslop was considered a scholarly dreamer and was well-known locally for his innovative ideas. He spoke frequently at farming institutes and seminars, but his crops were notoriously poor. He died in 1920, deeply in debt. In 1931, his widow sold the

PRESERVATIONISTS have been working since 1991 to save the intriguing Star Barn on Highway 283 near High Spire, Pennsylvania. A star appears in each of the barn's four gables. Built by landowner and banker John Motter in 1872, the barn was large for that era, with three threshing floors and an extra-high ceiling. Preservationists hope to restore the barn and open an educational center there. Zelda Rowley of Mt. Rainier, Maryland provided the photo.

farm to the local Ford dealer. Unable to paint the barn during the Depression, he coated it with used motor oil from his dealership to preserve the wood, which turned it an unusual sienna color. A few years later, he repainted the barn white.

The barn burned down, possibly from a lightning strike, on July 9, 1985. All that remains is a skeleton of concrete foundations and watering troughs. But the remarkable gabled barn still lives in the memories of the many people who considered it "their" landmark. They remember it as the embodiment of a dream.

George Hyslop and his wife, Julia, are buried in a cemetery on the property, their graves facing West View Farm and what was once "the best barn in the land". —*Karen Sunderman*
Deshler, Ohio

This stone barn, which stands on our 320-acre family farm near Chokio in western Minnesota, was a 20-year labor of love for my grandfather, Frank Schott, and his sons, William (my father) and Tony.

The sturdy barn is 50 feet long and 33 feet wide and the stone-

DISTINCTIVE BARN in the northern Berkshires features an unusually shaped front section and elaborate cupola. **Paul Turnbull of Charlemont, Massachusetts took the photo.**

STURDY SCHOTT BARN in Stevens County, Minnesota was built almost entirely of fieldstone and concrete. Wood was used only in the roof and the door and window frames (left). Photos snapped by Kathleen Brandt of Jeffers.

and-concrete walls are 18 inches thick. The concrete loft floor, which is supported by iron I-beams, weighs about 75 tons.

Grandpa, a stone mason by trade, started building the barn in 1923. It took 20 years to complete because work was often interrupted by construction jobs and farm duties.

Rocks for the walls were gathered from the fields and dragged by horses and a 10-20 Farmall tractor. Dad says one boulder was so big it took four horses and a tractor to move it into position!

Using horse-drawn wagons, Dad and Uncle Tony hauled sand for the concrete from Lake Hattie, 5 miles away. The floor, feed bunks and the curving loft stairway are all made of concrete and remain smooth and uncracked to this day—a tribute to my grandpa's skill and craftsmanship. —*Bill Schott Jr., Chokio, Minnesota*

UNCUT BEAMS provide the center—and only—supports for this striking cantilevered barn (below) in Cades Cove, Tennessee. Patricia Bridges of Lexington, Georgia says the barn dates to the early 1800's. "It was built with no outer supports so the owner could get his animals, buggies and wagons out of the weather with nothing in his way," she explains.

DAIRY BARN (above) near Chassell, Michigan was originally part of a Finnish homestead along the Sturgeon River. The owners, Dan and Lisa Adams, of Murphy, North Carolina, bought the barn, which was built in 1936, from the homesteaders' descendants in 1995. The lower-level walls were made of 1-foot-long tamarack logs stacked sideways in cement. For the roof, 1-inch tamarack boards were layered four deep, then covered with tin. Lengths of wood set in concrete provided a solid base for such "stove wood barns". Inset photo, from Sandra Taylor of Elsie, Michigan, shows log ends up close.

A SILO inside a barn makes Bill Petersen's barn near Birchwood, Wisconsin most unusual. "Maybe the farmer that built it wanted to keep the silage from freezing," he observes. "We get some cold weather up here in northwest Wisconsin."

"THE UMBRELLA BARN" is what natives of Wood County, Ohio call this unusual structure, which is two side-by-side barns joined by a round roof. Leanne Wolff-Langenderfer of Tiffin says so-called "twin barns" are common in the Black Swamp area of northwestern Ohio. "They're testaments to the growth of agriculture at the turn of the century."

YELLOW ROOF marks this Washington barn as out of the ordinary, although the tidy-looking brick walls make it distinctive, too. Richard Gann of Eaton, Colorado took the photo.

BRICK-END BARNS are rarely seen outside central and southeastern Pennsylvania, barn historians say. This one, owned by Ted and Loretta Pesano of Dillsburg, features lily, diamond and hourglass designs for ventilation (the patterns are created by leaving out bricks during construction). "Structurally, the barn is sound," reports Loretta (that's her above). "Such structures would cost a small fortune to build today, but in the mid-1800's, when this one was built, they were actually less expensive than wooden barns."

BARN spotted by Jerry Irwin outside Manchester, Michigan has some rounded edges. Jerry hails from Paradise, Pennsylvania.

LANDMARK stone barn (left) in Chase, Wisconsin was built in 1903 and restored in 1995. The owners are brothers Casimir and Stanley Frysh. Neighbor Jo Ann Wilcox, who provided the photos, says the whole community is grateful for their efforts to save this rare structure, which the Wilcoxes use for machinery storage.

THERE'S a good reason Ernest Olson of Cavalier, North Dakota is so interested in this twin barn near his home—his grandfather helped build it back in 1914. The barn measures 68 feet by 70 feet. The left side was used for horses, and the right side for cows and calves, Ernest relates.

HORSE BARN on Donamire Farm in Lexington, Kentucky is majestic enough to be mistaken for a church. Becky Croslin of Dallas, Texas says the barn is made from the limestone that's so plentiful in Bluegrass Country.

77

On the farm in Utah where I grew up, there was a large barn that always had pigeons in its loft. One day, I decided I'd like to capture a baby pigeon, so I climbed up to the loft, even though my father had forbidden me to go there.

As I reached for a pigeon, the rest all flew away, which frightened me so much that I fell out of the loft. Luckily, I fell in some soft hay.

My father ran over and said, "Are you hurt, son?" I answered shakily, "N-n-no." When he found that I was all right, he turned me over his knee and spanked me. I never climbed up in the loft again!

Today, 60 years later, the barn still stands, although part of the roof collapsed recently. It's still a favorite subject for artists.

Francis Tate built the barn from red pines believed to be taken from the Snake Creek Canyon. It was a timber-frame barn built with mortise-and-tenon construction and wooden dowels. It was the first of its kind in Wasatch County and is now part of a state park.

—Andrew Besendorfer, Midway, Utah

ALTHOUGH weathered, the Tate Barn in Utah remains a commanding presence. Andrew Besendorfer, who grew up on this farm, says the barn is especially popular with local artists.

LIMESTONE BARN near Manhattan, Kansas is still very sturdy, reports Dorothy Dean Miller of Manhattan, who took this photo.

WHEN Otego, New York farmer Sherman Burdick hosted a barn-raising in 1894, he insisted on having the tallest barn around—and he got it. His four-story marvel, with twin two-story hay-mows, was a local landmark until its collapse during a winter storm in the late 1980's. Margaret Gladstone of Otego grew up on the farm and still misses the old barn. "My siblings, friends and I have wonderful memories of playing there," she says.

SPARKLING concrete-and-steel barn in Lexington, Indiana was built by the William English family in 1914 to house farm animals and a modern dairy. A large group of laborers built the three-story barn in 100 days—at a cost of $75,000. Now part of Englishton Park, operated by Presbyterian Ministries, it's used for summer camps and retreats, says Doris Prewitt of Lexington.

WINDMILL atop this barn on the Brandel Farm near Stockholm, Minnesota is used to grind grain as well as provide water for cattle, explains Connie Cleaver, Nampa, Idaho.

EVERYONE around Superior, Nebraska is familiar with the "the twin barn place", where horses were raised to sell to the Army during World War I. The photo below was taken during a horse sale in 1915. Current resident Marsha Ray says the two complete barns are connected by a granary. "They're still in use today, but I'm sad to say time has taken its toll, and these old barns are about to become only a memory," she reports. The photo at right was taken in 1978.

APPEARANCES can be deceiving, as Douglas and Lee Roberts discovered after buying a small farm in Erie, Illinois. Their "barn" is actually an old railroad boxcar!

When my son and daughter-in-law bought a small farm in our town, they knew the place had a barn. Only later did they learn the "barn" had once been a Rock Island Railroad boxcar!

No one seems to know how the original owner got the boxcar to the farm, but neighbors recall he used it for storage and later to house cattle. As time passed, he built onto the boxcar a little at a time until it looked just like any other barn.

But if you look closely, it's clear this barn is different. It sits on a raised cement slab and still has the original couplers, the hardware for the ladder that led to the roof and the original sliding door.

When you step into the barn, it's obvious you're standing inside an old boxcar. Although the car's roof has been removed, the metal struts remain. The barn is a real conversation piece for everyone who visits. *—Marjorie Roberts, Erie, Illinois*

BREWER Ernest Klinkert built this majestic barn near Racine, Wisconsin to house his racehorses in the early 1900's. "That explains the unique turrets, fancy trim and wrought-iron racehorse weather vanes," says Mary Piper of Sterling, Virginia, whose in-laws now own the farm. Mary adds that artists often stop at the house to ask if they can paint or draw this landmark.

WEATHERED SIDING and arched stone doorways add to the appeal of this handsome barn near Wilkes-Barre, Pennsylvania. Lois Hamilton of nearby Doylestown provided the photo.

STUCCO BARN is one of the few left in southern Iowa and is still used today, writes owner Greta Bata of Des Moines. Located east of Plano and west of Centerville, the barn took about 3 months to build in 1948. The builders, including Greta's dad, worked atop scaffolding for 6 days.

RARE Pennsylvania Dutch barn in Berks County, Pennsylvania features hex signs, "devil's arches" and decorative brick vents. Michael Gadomski of Sterling took the photo.

"WE BOUGHT our farm in 1991, mainly because of this big old barn; we fell in love with it," writes Stuart Vencill of Joseph, Oregon. "It was built in 1916, and each wing is 78 feet long and 36 feet wide. It contains nearly 5,000 square feet of space on each floor and the hayloft holds 300 tons of loose hay. We're fairly sure it's one of the biggest barns in the state. We dream of holding dances someday in that huge hayloft!"

CANDY-STRIPED BARN is a favorite of Agnes Hochstein's grandchildren, who make a game of looking for it whenever they drive from her home in St. Helena, Nebraska to visit relatives in North Bend. "They know that once they see the barn, North Bend is very close," Agnes says. "We all love this cheery-looking barn."

SPECTACULAR BARN (right) at Old Fields, West Virginia was the largest wooden structure in the state when completed in 1917 for farmer George Leatherman. Passersby often mistook the medieval-looking building for a church. Today, the "fairy-tale barn" is in danger of collapsing and the family has applied for preservation funds to save it. Marjorie Franklin of Harrisonburg, Virginia shared the photo.

FORMER STAGECOACH STOP north of Billings, Montana looks much the same in 1995 (below) as it did in 1905 (at left). Built in the 1880's, the 30-Mile Stage Station provided service to stagecoaches traveling between the Yellowstone River and Fort Benton on the Missouri River, explains Dianne Cooper of Billings, who shared the photos. Dianne says her husband's family operated this stop for many years.

ON SUNNY DAYS, the gleaming roof of Charles and Beatta Robben's barn at Grinnell, Kansas is visible from Interstate 70, 7-1/2 miles away. Local folks refer to it as "the big white barn with the silver roof", Beatta says. Built in the 1920's, the barn is as tall as it is long—30 feet—and 56 feet wide.

BARN near Maple Plain, Minnesota receives some support from twin silos that flank its sides. Connie Cleaver of Nampa, Idaho photographed the barn, which is currently being restored.

"HOP HOUSES" were once a fixture on virtually every farm in Otsego County, New York, native Niles Eggleston writes. Today, only two or three remain. This barn, built before the Civil War, stands near Burlington Flats.

H ops were once *the* crop in Otsego County, New York, where I grew up. Our county was the largest producer of hops in the state—and our state led the entire nation in hops production from 1830 through 1910.

For two generations, our entire economy revolved around the production of hops. At harvesttime, schools and businesses closed as teachers and students, business owners and employees worked side by side in the fields. Harvest was a social event, too. Many young people met their future spouses while plucking the aromatic blooms.

The blossoms were taken to barns known as "hop houses", or hop kilns. A furnace on the first floor dried the crop in the curing room above. A characteristic cupola near the end of the barn served as a ventilator and made hop houses easy to identify.

Prohibition, blue mold disease and other factors had decimated hops production by the time I was a young boy. But practically every farm in the county still had a hop house, even though it was being used for something else. There were hundreds and hundreds of these barns.

Today, no more than two or three remain. The most beautiful is a petite building (above) on the Osborn Farm on Jacobs Road, near Burlington Flats. Built before the Civil War, it's struggling to survive the elements. But to a handful of *human* survivors, it still brings back memories of the joys of a simpler time.

—*Niles Eggleston, Milford, New York*

OLD BARN (shown at left in 1930 and below in 1972) on Ed Strobel's farm in LaSalle County, Illinois was reluctantly dismantled after high winds tore off the lean-to on the right. The construction year, 1873, was painted in the circle in the upper gable. That's Ed standing on hayrack in photo at left.

The barn on our farm just east of Ransom, Illinois withstood the ravages of more than 120 years of weather, violent storms and hard use. In fact, it outlived many of our outbuildings, along with three houses!

Designed by an architect, the barn was built in 1873 for about $650. A German-style bank barn was an unusual sight on the flat prairies of Illinois, but this one was even more unusual because it had an oak bridge leading from the bank to the loft, with a passageway below.

It was wild at haying and threshing time. Some of the horses were afraid to go onto the floor because their hooves made a terrible racket.

It also became very dark at times because entering wagons would block the light from the rear door. And when horses saw light coming through the sliding doors at the opposite end of the barn, they'd head straight for it—and a 10-foot drop to the ground below. A huge plank kept them from moving forward, but at least one team went right on through. Luckily, a manure pile softened their fall.

After unloading, we had another problem—backing the teams down and out of the loft. Unruly horses always had a bad time of it, and we seemed to have a few of those.

Like most barns of the era, this one was put together completely on the ground, then pulled up by horses. The square nails were handmade by a blacksmith. Most of those rusted out, and the barn was completely re-nailed twice. The original roof lasted over 65 years and was replaced only once, with corrugated steel.

In 1993, strong winds destroyed the lean-to for cattle on the barn's west side, leaving it open to the elements. The rest of the barn was still structurally sound, but there was no way to close up the hole. The barn had to come down.

I dismantled it in 1995 with our tenant and his family, who helped salvage all the good beams and heavy joists. It was a sad day when we started, and even sadder when we finished.

Now all that remains is a concrete horse tank, a cement floor and retaining wall and a road—leading up to nowhere.

—Ed Strobel, Decatur, Illinois

CIVIL WAR-ERA horse barn concealed a well at the foot of the stairs, reports Ann Culvahouse of Rockwood, Tennessee. When troops passed through, R.H. Johnson watered his Arabians in the barn, where the soldiers couldn't spot them. The barn is on Highway 58 near Ten Mile, Tennessee.

"THIS BARN must be one of a kind, because I've never seen another one with a stream under it!" writes Jane Johnson of Chambersburg, Pennsylvania. Her parents bought the property in nearby Center in 1947 but don't know how old the barn is—or why it was built over water. Her father laid stone walls to prevent flooding. "Growing up here, I often played in the stream under the barn," Jane recalls. "Now, how many people can say that?"

MATERIALS used to build this barn in Mayville, Wisconsin make it very unusual. LeRoy Schultz of Morgantown, West Virginia says the entire structure is made from railroad ties.

The three-story barn (right) on our ranch near East Helena, Montana has 70 windows, stands 54 feet tall and contains *27,000 square feet* of floor space!

William Child had it built in 1888 after most of his cattle froze to death on the open range during the harsh winter of 1885-86. He wanted a barn big enough to hold 500 cattle and 300 tons of hay—enough to feed a herd through even the worst winter.

The place has been in my family for 52 years and remains a working ranch. Cattle are housed on the barn's first floor, and we still use the third-floor haymow. The second floor is used to store grain, machinery and a few antiques.

Our ranch is on the National Register of Historic Places, and we've offered tours since 1994.

Our octagonal stone house is unusual, too. Built around the same time as the barn, it's believed to be the oldest octagonal house west of the Mississippi.
—*Paul Kleffner*
East Helena, Montana

THIRTY stone masons were hired to lay the glacial fieldstones that flank the wooden center section of Paul Kleffner's barn. Rancher William Child had the barn built in 1888, after most of his cattle froze to death on the open range. Jim Gipe took the photo.

"THIS is one of the Welsh barns owned by friends of mine, Chris and Martha Young of Leroysville, Pennsylvania," relates Roger Kingsley of Columbia Cross Roads. "Isn't it a beauty?"

FARMER Bert Kennell used stones collected on his farm to build this barn near Reeds, Missouri in the late 1920's or early '30's. "They say the only sure crop in Missouri is a rock crop," chuckles Polly Jackson of Enterprise, Mississippi, who took the photo. The farm's current owner is Ivan Cunningham.

UNUSUAL HOP BARNS like this (above) have all but disappeared from Washington's Puyallup Valley, reports Bill Creelman of Bremerton. "Most of them were located near the small town of Orting," he recalls. Bill also snapped the photo at right of a unique barn near Port Gamble.

K ari and Loren Schwinghammer's unusual barn south of Cologne, Minnesota combines a gothic-roofed rectangular barn with a domed round one.

William Lemke, a prominent dairy farmer, built the barn on his Lonely Oak Farm in 1927. The rectangular section measures 34 by 48 feet and includes horse stalls and bull and calf pens.

The round section is 60 feet in diameter, with a center silo that holds the dome's supporting rafters. Thirty-six cow stalls surround the silo for convenient feeding.

A double row of ornate concrete blocks forms the 16-inch-thick walls. The original cedar shingles, though in need of replacement, still cover the roof. Each shingle was hand-tapered with a knife to accommodate the curve of the dome.

—*Joseph Koenig*
Belle Plaine, Minnesota

MINNESOTA dairy farmer William Lemke built this concrete block barn in 1927. It still has the original cedar shingles, which were shaped by hand to fit the curve of the dome.

"I FOUND this old abandoned potato barn, or potato house, in the Presque Isle area of Maine," reports Shari Ireland of Mapleton. "Note the unusually low roofline. The barn provides landscape painters and photographers with plenty of material."

DECORATIVE CUPOLAS set this barn apart from the crowd. Jerry Irwin of Paradise, Pennsylvania spotted it along Route 22 near Huntingdon.

For weeks in the early 1900's, my dad, John Noble Machey, spent all his evenings with a flat carpenter's pencil and large sheets of brown wrapping paper, drawing plans for his new barn. It was different from anything that had ever been built before.

"For half my life, I've pitched hay up into the mow," he said. "I want to spend the other half pitching it *down* into the mow."

When a barn builder looked at the plans, he wanted nothing to do with it. He said he wouldn't risk his reputation on such a monstrosity. All the other carpenters had the same reaction. Why build a barn on a steep hillside? Didn't Dad have plenty of level land?

In the end, Dad built the barn his way, and it was a honey. This versatile barn eliminated all outbuildings. The wagon shed, pigpen, sheep shed, henhouse and straw stack—even the barnyard—were all under one roof.

The driveway floor, where all the threshing was done, sat above the hay and grain mows. Wheat and oats were stored in large bins under the floor, with openings for feeding in the runway between the horse stalls and cow stanchions.

We had one old mare who knew how to use her teeth to pull

open the sliding valve on the oat bin. Sometimes we'd find her with such a violent bellyache we had to call a horse doctor!

The basement was like a giant three-level staircase, with sheep on the top "step", horses and cows on the second, and pigpens, a watertight cement-lined manure pit and a barnyard on the lowest level.

Troughs in the basement were filled by the simple turn of a faucet. The water came from a large cistern on the upper level.

Dad took great pride in his barn, and enjoyed showing skeptical farmers around. During one tour, he walked up to a trough, put his hand on the faucet, and said, "I let gravity do my work for me."

Just then, everyone noticed pigs wading and swimming in hundreds of barrels of water in the manure pit. When the livestock were watered the night before, someone had forgotten to turn off the faucet!

—*Written by James Dean Machey*
and submitted by his niece,
Frances Parlin, Marietta, Ohio

WHEN he couldn't find anyone who'd build the unique barn he designed, John Noble Machey did it himself. "The barn is still standing in Washington County, Ohio," says his granddaughter, Frances Parlin.

"THERE are very few of these kind of barns in North Dakota, so this one is special," asserts Barb Pelzl of Fargo, who took the photo above. "Notice how low the roof is to the ground. I love this barn!"

NOT your run-of-the-mill red barn, this beauty stands outside of Mora, Minnesota, reports Connie Cleaver of Nampa, Idaho.

CLASSIC Pennsylvania Dutch barn, located near Kutztown in eastern Pennsylvania, features cozy closed forebay, well-maintained hex signs and so-called "devil's arches". According to folklore, hex signs ward off evil or ensure good fortune, and the white arches repel devils and witches. Jerry Irwin of Paradise snapped the photo.

"FLOODING every spring forced this farmer to build his barn in western Washington on stilts," writes Mrs. Weldon Neuschwanger of Olympia. "Bet you've never seen a barn like this before!"

STONE-AND-BRICK foundation on above barn, located east of Hustisford, Wisconsin on Highway 109, was snapped by JoAn Wolter of Horicon. "Notice the 'eyebrow' trim above the windows and doors," she says. JoAn also photographed more fine brickwork on a barn north of Hustisford on Highway E. "Note the cross over the door (right), asking for blessings on all the work done there."

HANDSOME stonework, immense size and covered passageway from ramp into barn make this Pennsylvania beauty near Farragut most unusual. Photo was taken by Jerry Irwin of Paradise.

UKRAINIAN immigrants Jakob and Sophia Sauter followed old-country architectural style for this barn-like building on their homestead northwest of Carson, North Dakota. The stone structure, built in 1903, had three separate sections—living quarters in front, a blacksmith shop in the middle and a barn in the rear. The building is 18 feet wide and 80 feet long, says descendant Edwin Sauter, who still farms his ancestors' land.

STATELY BARN in New York's Finger Lakes area (right), near Conesus, was discovered by Rose Harshbarger of Rochester. She doesn't know the barn's age but notes that the turret resembles the "widow's walk" that was popular on homes around 1850.

MAGNIFICENT BARN (left) on the East Rainy Butte Ranch, just south of New England, North Dakota, was built in 1915, explains Shelley Heidegger of Yacolt, Washington. "The man who built it owned a lumberyard. My husband's family bought the 1,120-acre ranch in 1928. It's since been sold, and the great old barn is no more."

90

STATELY LADD BARN (left), built in 1906, sits atop a knoll west of Sherman, Texas. The barn was south of Sherman for years—until a descendant of the original owner gifted it to Betsey Spears for $1. She had the barn dismantled and moved in seven pieces to this site, where it was carefully reassembled. Says Ron Wheeler of Bonham, Texas, who shared the photo, "It looks very much at home surrounded by native bluestem pastures, grazing cattle and seasonal wildflowers."

STONE BARN on the El-Jireh Farm (right), located on Highway 21 near East Jordan, Michigan, was snapped by Lawrence Lake of Lennon. "I don't know the history of this barn, but from the looks of it, the stone mason knew what he was doing," Lawrence observes.

ARCHED DOORWAYS of handsome feeding barn (below) near Ponca City, Oklahoma allow livestock to find shelter quickly during bad weather. The barn, made of native stone, is nestled into a slope so hay and feed can be driven into the upper level, then unloaded and dropped to the animals below. Lewis Gilbreath of Winfield, Kansas shared the photo.

Chapter 5

Round Barn Roundup

Why Build a Round Barn?

Barn lovers are particularly fond of round barns (also known as multi-sided or polygonal barns), those eye-catching novelties popularized in the late 1800's.

At the time, innovative farmers figured circular or polygonal barns would require less lumber, as well as better withstand high winds. Others said they could be built faster. Yet others reasoned that since cows are narrower in front, it would be more efficient to have them feed around a central silo—arranged much like wedges of pie in a pie tin—instead of standing in rectangular stalls. Superstitious folks had their reasons, too—no corners meant no place for evil spirits to lurk!

Their ideas caught on. Historians estimate there were more than 1,000 round barns at the height of their popularity, most in the Midwest. But over the years, more than half burned down or were abandoned, mainly because their shape hampered expansion and new mechanized, bigger equipment didn't work well in them.

Fortunately, many have been restored, ensuring that these monuments to ingenuity and rural pride remain standing tall and proud.

BROWN SWISS DAIRY BARN (above) near Shipshewana, Indiana is made mostly of concrete. Dairy farmer Menno Yoder and his sons poured the concrete into handmade forms when they built the structure in 1907. Richard Gann of Eaton, Colorado took the photo.

FALL LEAVES and grazing horses make the round barn above near Royalton in Waupaca County, Wisconsin even more beautiful. Darryl Beers of Algoma snapped the photo.

PATTY SIMON of Hoven, South Dakota discovered the interesting-looking barn at left near her home a couple of years ago. "I was intrigued by its architecture and weathered look," she relates. "Imagine my surprise when I learned that it's owned by my brother-in-law!"

SPACIOUS CUPOLA provided ample ventilation for this barn in Buffalo County, Wisconsin. Sandra Ebert of nearby Alma says "1913" was scratched into the cement, so local residents assume the barn was built that year. Agricultural engineers at the University of Wisconsin were promoting construction of round barns from 1900 to 1915, Sandra says.

TWIN SILOS make this three-tiered round barn look like a rural castle, says Dorothy Dean Miller of Manhattan, Kansas. Built in 1913 near Harper, the barn is 80 feet in diameter and stood 65 feet high from floor to peak—until a windstorm knocked off the top cupola several years ago.

In the early 1900's, a Chicago brain surgeon named Lawrence Ryan was looking for a weekend retreat and a home for his small herd of Black Angus cattle. His search ended when he found 80 acres just off what is now Route 78 in Kewanee, Illinois.

For his barn, Dr. Ryan chose a round design first used in the United States by the Shakers in the early 1800's. These barns were considered tornado-proof, and the circular feeding and waste removal systems were space- and labor-efficient. They had one other advantage—tax collectors often underestimated their size!

A crew of Southern carpenters who specialized in round barns completed the construction in 1910. The cost was a whopping $9,600—nearly four times the cost of the average round barn of that period.

The three-level Ryan barn is 85 feet in diameter and 80 feet high. The 16-foot white pine boards used for the curved horizontal siding were soaked overnight in a lake to make them bendable, then hammered onto the frame wet.

The site became part of Johnson Sauk Trail Park in 1969. Five years later, amid rumors that the state might raze the deteriorating barn, citizens rallied to save it. Thanks to them, the barn was listed on the National Register of Historic Places in 1974; restoration efforts followed.

Fewer than 50 round barns still exist in Illinois. Some have

ANTIQUE FARM MUSEUM is housed in Ryan's Round Barn at left, now part of Johnson Sauk Trail Park in Kewanee, Illinois. The intricate web of woodwork in the ceiling, shown below, testifies to the carpenters' craftsmanship. Guided tours are given May through October on the first and third Sundays of each month.

been lost to vandals, others to the effects of time. But thanks to a dedicated group of citizens, this particular piece of America's past has been preserved for posterity.

—*Susan Wildemuth*
Atkinson, Illinois

In 1889, a horse foaled in Montana and named "Spokane" won the Kentucky Derby. That proved to be an extraordinary event, since no horse from Montana has won the Derby since then.

The barn where Spokane was born is just as extraordinary. Located at the base of the Ruby Mountains, near Twin Bridges, the three-tiered barn was designed and built by Noah Armstrong, a champion horse breeder and the head of a gold-mining company.

When finished in the 1870's, his barn was part of the largest breeding layout west of the Mississippi. Its unique design includes a spiral staircase and mortise-and-tenon joints. The stone abutments stand atop piles driven 12 to 15 feet below water level, down to bedrock. The walls are three boards thick, separated by double layers of building paper.

The first floor measures 100 feet in diameter; the second floor, 76 feet; and the third, 36 feet. The outside compartments on the ground floor held offices, tack closets, two hospital rooms, employee dormitories and 12-foot-square box stalls that each opened onto its own 2-1/2-acre paddock.

A 20-foot-wide, 1/8-mile track allowed employees to exercise horses during bad weather. In an emergency, the track could shelter as many as 300 horses. The main entrance to the track was wide enough to accommodate a 10-horse wagon loaded with supplies.

The second floor held hay and grain, and the third contained a reservoir of 11,000 gallons of water, supplied by a windmill atop the barn. This arrangement allowed water to be funneled directly to each stall and provided a ready source of water in case of fire.

A 4-by-8-foot wood carving by Mr. Armstrong's daughter, Emma, hangs over the main entry. The carving is based on Rosa Bonheur's famous painting, "The Horse Fair". Although weathered from a century's exposure to the elements, the details are still visible.

—*Alice Schumacher, Great Falls, Montana*

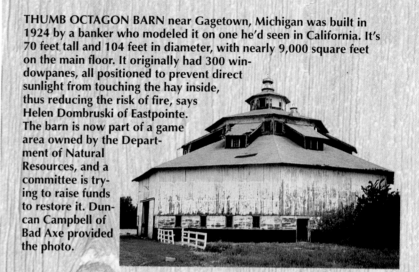

THUMB OCTAGON BARN near Gagetown, Michigan was built in 1924 by a banker who modeled it on one he'd seen in California. It's 70 feet tall and 104 feet in diameter, with nearly 9,000 square feet on the main floor. It originally had 300 windowpanes, all positioned to prevent direct sunlight from touching the hay inside, thus reducing the risk of fire, says Helen Dombruski of Eastpointe. The barn is now part of a game area owned by the Department of Natural Resources, and a committee is trying to raise funds to restore it. Duncan Campbell of Bad Axe provided the photo.

HORSE BREEDER Noah Armstrong built the unique barn above in the 1870's, providing the foaling place for Montana's first and only Kentucky Derby winner. "Spokane" won the race in 1889, the same year Montana was admitted to the Union.

BREATHTAKING BARN was photographed in Arelee, Saskatchewan, about 50 miles south of Saskatoon. William Corbett of Central, Utah, who shared the photo, notes the cattle stalls are arranged facing the center of the barn, where hay is dropped down from a loft.

BUILT in 1910 in Carroll County, Indiana, barn above is owned by Jerry Britton. Oral Wright of Middletown says Indiana still has about 105 round and polygonal barns, but the number diminishes every year.

BLUE BARNS aren't very common—particularly round ones! The one pictured above is in Iowa.

PLEASING hexagonal barn at left was built by Barbra Martin's great-great-grand-father, Kannon Gilmore, between Willard and Ash Grove, Missouri, probably in the late 1800's. "When my mother was little, family get-togethers were held at the farm," says Barbra, of Rancho Santa Margarita, California. "All the little ones had such fun climbing into the loft and jumping down into the arms of their older cousins."

Our barn wasn't built to hold large machines, Percheron horses or lots of grain. It was built for the small things you need on a farm—tools, a few bales of hay, small pieces of machinery, a place for children to play. It was built round so the devil couldn't find a corner to hide!

My great-uncle, Harry Price (that's him at left), had the barn built west of White Hall, Illinois in 1912. The 6-inch-thick glazed tile blocks were fired by a local sewer pipe and stoneware company.

With a diameter of 36-1/2 feet, it's much smaller than the typical round barn—usually about 60 feet. Ours isn't the state's smallest, though—another in Clay County is 30 feet in diameter. Historians consider them significant, because they prove round barns were being promoted for smaller operations as well as larger ones.

I've recently restored the barn as a tribute to my great-uncle's progressive spirit. It's a featured stop on the annual Greene County Days Barn Tour.
—Robert Price, White Hall, Illinois

"MY UNCLE, Robert E. Lee Johnston, built this barn (right) near Red Bluff, California in 1911 to train horses for the Army during World War I," says Betty Nichols of Lotus, California. "The horses could run without getting into corners, making them easier to break." The Fort Crook Museum in Fall River, California is patterned after this barn, Betty says.

RIDERS on the "P" Ranch near Burns, Oregon used round barn below to train and exercise horses in winter. Built by ranch foreman Peter French around 1880, it has a central stone corral 60 feet in diameter writes photographer Diane Gill of Mount Vernon. Old-timers say the roof's support system (below right) was inspired by the ranch owner's wife's parasol. Charles Hale of Crescent took the interior photo.

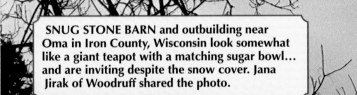

SNUG STONE BARN and outbuilding near Oma in Iron County, Wisconsin look somewhat like a giant teapot with a matching sugar bowl… and are inviting despite the snow cover. Jana Jirak of Woodruff shared the photo.

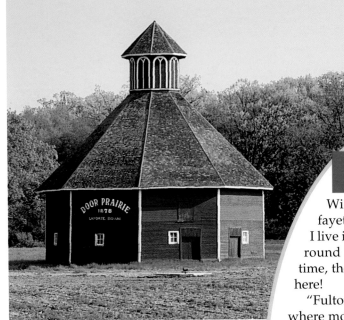

DOOR PRAIRIE BARN in Pleasant Township, La Porte County was the second circular barn built in Indiana. It was built in 1878.

CENTRAL WOOD-STAVE SILO helps support the roof of the Kellerman Barn near Romney in Tippecanoe County. The 60-foot-diameter barn was originally built for sheltering cattle and horses.

I have always been fascinated with round barns," notes Marsha Williamson Mohr of West La-fayette, Indiana. "I'm lucky because I live in Indiana, which is known as the round barn capital of the world. At one time, there were 226 round barns standing here!

"Fulton County (in north-central Indiana), where most of these photographs were taken, contains more round barns than any other county in the state. It even holds an annual round barn festival.

"I can't explain my infatuation with round barns—they just excite me. Why, I remember the day I went to photograph Indiana's largest round barn. When my girlfriend and I first saw the roofline in the distance, you'd have thought we were the first pioneers to see the Grand Canyon!

"Unfortunately, many of these barns are being torn down. As a result, I feel obligated to photograph as many as I can so future generations can ap-preciate them as much as I do."

MAX HOFFMAN BARN in Newcastle Township, Fulton County was built in 1911. The 63-foot-diameter barn can hold 5,000 bales of hay.

CLENDENNING BARN in Clinton County was built in 1912 for J. Seymour Clendenning. It's 66 feet in diameter and 70 feet to the top of the cupola.

BOARDS for the Utter-Gerig Barn above, built in 1915 in Henry Township, Fulton County, were soaked in a nearby creek to make them pliable. The bank barn has a stone foundation and three-pitch gambrel roof.

BUILT IN 1914, the Haimbaugh Barn above in Rochester Township, Fulton County features a 20-foot-wide shed built halfway around the barn. Just west of the barn, a treaty with the Potawatomi Indians was signed in 1836 on the banks of Chippewanuck Creek.

WILBUR BREEKS BARN at right closely resembles the Chicago Wrecking House Company's "Round Barn Design #206", which was advertised in the *Indiana Farmer* and other agricultural newspapers between 1910 and 1915. It was built in Union Township, Montgomery County in 1912-13.

AFTER being moved and reassembled in 1991 near Rochester, the Larry Paxton Barn became part of the Fulton County Museum complex. The barn was built in 1924 and measures 60 feet across.

I had heard of "the round barn on Route 66" for years, but never visited it until we learned our neighbor, Luke Robison, was restoring it. When we finally went for a look, I was amazed. I've never seen anything that compares.

The barn is in Arcadia, about 19 miles northeast of Oklahoma City. It was built in 1898 by Oklahoma Territory pioneer W.H. Odor, who wanted nothing but the best—in this case, a round barn, which supposedly could better withstand tornadoes.

The barn took almost 6 months to build. The rafters were green burr oak, cut at a sawmill on the farm. The timber was soaked in water until it was soft enough to bend.

To finish the roof, two rafters were tied into a peak at the top, and a ladder was built from the loft floor to the top and braced. None of Mr. Odor's workers were willing to climb up to tie the rafters, so he had to do it himself! There is no traditional bracing; each piece of lumber acts as a brace.

The barn is 60 feet wide and 43 feet tall. Mr. Odor compared its acoustics to those at the Mormon Tabernacle in Salt Lake City, Utah. It's been said that when the barn was new, you could stand near the wall and hear a pin dropped on the opposite side.

Over the years, the barn developed a large "twist" and was in danger of collapsing. Mr. Odor's heirs wanted to donate the barn to a group that would save it. The Arcadia Historical and Preservation Society formed for that purpose in March 1988. Three months later, the roof collapsed.

Luke (in the photo at right), a retired carpenter, wanted to help save the barn. He'd never built one, but he has an old-fashioned approach to craftsmanship and a strong feeling for history. He relished the challenge of discovering how this architectural wonder had been built with the limited tools available in the late 1800's.

Luke and a group of retirees who called themselves "The Over-the-Hill Gang" began clearing debris from the fallen roof. Over the

OKLAHOMA has only one round barn, so there was great interest in saving the beauty near Arcadia. The Arcadia Historical and Preservation Society collected about $40,000 in donations for the project. Many of the materials and virtually all the labor were donated. Edie Greenfield of Luther provided all the photos except the one at right, taken by Jerry Irwin of Paradise, Pennsylvania.

next 4 years, these seniors and an army of volunteers realigned and stabilized the walls, replaced deteriorating siding and installed a new roof.

The importance of the volunteer effort can't be overstated. Contractors estimated the project should have cost $175,000, but so much labor and materials were donated that the cost was kept to $40,000.

The restored barn was dedicated April 4, 1992. Now open to the public from 1 to 6 p.m. Tuesday through Sunday, it's had visitors from every state and 20 foreign countries. It has been used for many meetings, parties, weddings and dinners.

This barn is a monument to the workmanship, intelligence and perseverance of our area's early-day settlers. Thanks to Luke and the others who contributed so much, that monument will stand for generations to come.   —*Delores Bendure*
Oklahoma City, Oklahoma

"NOAH'S FOLLY" is what skeptical neighbors called this round barn when Noah Sheely hired a carpenter to build it in the early 1800's. But his grand-niece, Jeanne Martz of Orlando, Florida, says it's as sturdy today as when it was first built. Located off Route 30 near Cashtown, Pennsylvania, it's now the home of the Round Barn Farm Market. "When I took this picture in 1995, the barn looked no different than when I played in it as a child," Jeanne says.

MAJESTIC tri-level barn is on Highway 75 near Durand, Illinois. "It was built around the turn of the century, and the owner believes it was developed by the University of Illinois," says photographer Steve Snyder of Freeport. "They thought this design would revolutionize the way America farmed—and they took it seriously. That barn will stand forever." The first level was used for horses, the second for dairy cattle and the third for hay storage. The central silo has a 12,000-bushel capacity.

"WE SPOTTED this 18-sided barn in a remote farming area in Oregon and couldn't resist driving to it for a closer look," writes Maxine Mock of Pomeroy, Washington. "The years have taken their toll, but it's obvious a lot of work went into planning and building this fascinating barn."

CONCRETE BLOCKS were used by Henry and Martha Frantz to build this barn near Grand Junction, Iowa in 1911. Originally designed for beef cattle, it was converted to a dairy in 1930. "Today it's just a picturesque reminder of bygone days," writes neighbor Betty Bishop, who shared the photo.

"PLAYING BASEBALL inside his grandparents' barn every spring with his cousins and uncles remains my husband Royal's most pleasant memory of this barn," writes Jean Clausing of Jackson, Wisconsin. The barn, built in Mequon in 1848, was one of 14 octagonal barns just north of Milwaukee along Lake Michigan. The shape may have been chosen to help the barns withstand strong winds off the lake.

"THE STARKE BARN (above), located east of Red Cloud, Nebraska, is built into a gentle slope," writes Leota Huff of Commerce City, Colorado. "The central silo is made of brick, and stanchions and stalls encircle the outer edge. The upper levels have haymows and room for equipment storage. It's quite remarkable." The barn was built in 1902 without nails or pegs.

ROUND BARN at the Ojo Caliente Mineral Springs Spa Resort in Ojo Caliente, New Mexico was scheduled for restoration in 1996. The barn will house horses that guests will use to explore trails and Indian ruins in the surrounding hills, says Prita Shalizi of Sante Fe. The barn is listed on the National Register of Historic Places.

FOR HER 15th wedding anniversary, Phyllis Birney received quite a gift from husband Lawrence—this 16-sided barn near Mullinville, Kansas that she'd always loved! Phyllis later worked to get it listed on the National Register of Historic Places and donated it to the Kiowa County Historical Society, which is overseeing a $100,000 renovation. The barn was built in 1912, says Bob Neier of Wichita, who shared the photo.

My father, Harley Nelson, was a hired man on Theodore and Albert Ekstrom's dairy farm near North Branch, Minnesota. He was working inside their eight-sided barn (below) on September 11, 1942 when a tornado blew off part of the roof.

The repair crew had to call the original master carpenter, who'd supervised the barn's construction in 1917, to show them how to fix it. This proved no simple task.

The front half of the roof, where the peak and hayloft doors are located, was gone, and the back half had to be stabilized. To do this, the crew attached cables and pulleys to a neighbor's trees. They had to adjust them every day.

The storm also blew off some of the concrete blocks. Since they'd been made on the farm and would've been difficult to replace, the repair crew had to lower the entire roof by three blocks!

All this work was done without fancy equipment—just a lot of old-fashioned muscle power.

In 1954, my parents moved to this farm and kept a herd of Guernseys until 1976. The farm is inactive now, save for some rented cropland, but the barn remains a community landmark and is a favorite subject of photographers.
—*Jeffrey Nelson*
North Branch, Minnesota

IT COST $7,000 to build this barn of hollow clay blocks near Nashua, Iowa. Helen Debner moved to the farm with her family in 1945 and now lives in nearby Greene. "A windstorm damaged the roof beyond repair in 1994, and the barn was destroyed in 1995," Helen reports.

My parents acquired a round barn when they bought a farm in Perry County, Ohio in 1926. Built in 1917, it had a double slate hipped roof, a wooden silo in the center and a big cupola on top.

A fire that broke out during wheat threshing in 1932 destroyed the barn. That fall, a new barn was built on the same cement foundation. A son of the original barn's builder was foreman of the construction crew.

The new roof would be curved, so the rafters had to be curved as well. The green lumber was shaped on a form, then laminated in pieces over 100 feet long. Each piece had to be exactly the same length.

The men raised the rafters with a gin pole, fastening one end to a shelf atop the barn siding and the other to a laminated ring at the top of the silo. The rafters were held in place with 5-inch spikes.

The barn is 68 feet in diameter and about 50 feet tall, and it has a central cement silo 12 feet in diameter. The construction cost about $3,000.

Dad sold the farm a few years ago, but he always enjoyed working in the barn and hopes a few round barns will be preserved. It would be a shame for their beauty to be lost forever.
—*Alice Warne, West Liberty, Ohio*

WORKMEN used 1,800 pounds of nails to build this barn in Perry County, Ohio in 1932. All the lumber was cut on the farm, keeping the cost down to about $3,000.

UNUSUAL peaked roof on octagonal barn near North Branch, Minnesota was badly damaged by a tornado in 1942, posing tricky problems for the repair crew. To stabilized the back of the roof, cables and pulleys were attached to a neighbor's trees.

DAYBREAK provided a picturesque setting for the weathered "round" barn above in Brown County, Wisconsin. Photo was taken by Darryl Beers of Algoma.

ROUND cobblestone barn below, north of Bridgewater, New York, was used for drying hops before the Civil War, says James Hughes of Syracuse. Cobblestone was a popular construction material in central and western New York in the mid-1800's, he reports. The metal band encircling the barn may have provided additional support.

"THE RANCHER who built this round barn just wanted to be different, according to a local resident," writes Norma Stafne of Hettinger, North Dakota. She snapped this photo between Carter and Dutton, Montana.

It took 8 years to build my grandparents' barn (below) in Hancock County, Indiana, which was completed in 1903. It measures 102 feet in diameter and 100 feet to the peak of the cupola, making it the largest round barn in the state.

At haying time, 17 *rack wagons* could encircle the loft at the same time!

My parents moved to the farm in 1925. When my brother and I were old enough to start exploring the barn, we had strict instructions from Mother *not* to climb the stairs to the cupola. She didn't know it, but we climbed those steps every day, trying to get a little bit higher each time.

One day we finally made it to the top. The view from the cupola window was fabulous. We could see all our neighbors' farms for miles around.

While we were enjoying the view, we spotted Mother on the front porch and yelled at her. We thought she was going to have a heart attack!

Today the barn is owned by my brother, Don Kingen. He's kept it in excellent condition, but at considerable expense. Painting it in 1990 required 100 gallons of paint—including 20 gallons just for the white trim. Re-roofing has been a major problem through the years, as few workers are willing to tackle the job.

The barn has always been an attraction for local residents and tourists alike. Many want a closer look or ask to take photographs. My brother is happy to show them around and tell them the barn's history. We are all proud of our round barn. —*Carolyn Ramsey*
Mesa, Arizona

THREE GENERATIONS of the Kingen family have owned this barn in Hancock County, Indiana. As a child, Carolyn Ramsey and her brother secretly climbed the stairs to the cupola to savor the view of surrounding farmland.

For more than 15 years, Jerry Irwin of Paradise, Pennsylvania has been searching the countryside for barns to photograph—but none of them get his attention like round barns.

"It's their uniqueness that attracts me," says Jerry, whose photos are sprinkled throughout this book. "The best thing is going inside them and seeing the intricate carpentry.

"It's just magnificent—I can't get over it. Some of them look like a huge woven basket inside. It's either a carpenter's nightmare or dream—depending on the skill of the carpenter!"

These two pages feature some of Jerry's favorite round barns —enjoy!

ROUND BARN FARM outside Green, New York is aptly named. The farm is over 160 years old. The same crew that built this barn erected the brick beauty at left on Brookhaven Farm, about a mile down the road.

CATSKILL MOUNTAINS near Delhi, New York provide the setting for the weathered round barn above.

PRETTIEST round barn in Pennsylvania stands near State College, according to Jerry. This barn, built in 1910 from 100,000 board feet of white pine, gets photographed a lot!

TRUE round barn with ornate cupola adds rich color to this fall scene near Irasburg, Vermont.

STATELY red round barn overlooks the countryside near Rochester in northwestern Indiana.

MASSIVE round barn on Maple Avenue Farm near New Hampshire, Ohio in Auglaize County was built in 1908 by noted Midwestern round barn builder Horace Duncan for Jason Manchester. At 102 feet in diameter and 80 feet high, it is said to be the largest round barn east of the Mississippi River.

STEEP domed roof offers lots of storage space in this unusual round barn in the Palouse region of southeastern Washington, between Pullman and Moscow, Idaho. It was built in 1917 by farmer T.A. Leonard.

AFTER more than 80 years on the Aker family's farm near Bremen, Indiana, barn below was moved to Amish Acres in Nappannee. Michelle and Larry Zeiser, shown in photo at right with Michelle's grandmother, Marie Redman, posed at the barn's new site before their wedding reception. Michelle's great-great-grandmother was the barn's original owner.

M y grandmother was widowed in 1906, but that didn't stop her from building a barn. In 1911, Margaret Aker built a round barn on the family farm about 4 miles southwest of Bremen, Indiana.

My father, then 14, and his five brothers helped with the construction, along with a neighbor, Phillip Lauderman. The barn was made entirely of native hardwood. Mr. Lauderman used jigs to curve the green lumber as it dried.

The finished barn was 60 feet in diameter and 60 feet tall, with a cedar shingle roof. A straw shed added in 1930 was assembled with wooden pins, just as the barn had been.

My three siblings and I grew up on the farm, and the barn bore my father's name for about 60 years. The farm later was run by my brother, Frank Aker Jr.

In 1992, my brother could no longer keep up the place, so he sold it to B.J. and Shelley Schrome. They, in turn, donated the barn to Amish Acres, an 80-acre historic farm in Nappannee.

It took skilled Amish carpenters 1-1/2 years to dismantle the barn, mark every beam and rafter and move it to its new site, which includes other historic buildings the Amish have preserved.

The barn is now used as a theater. The haymow was reconstructed to seat 350 people and the straw shed is now the stage. It's beautiful and magnificent, and I'm proud that it's there.

I'm also proud to be a part-time tour guide at Amish Acres, telling others about "my" round barn. —Marie Redman Bremen, Indiana

Y ou won't find a listing for the world's largest round barn in the *Guinness Book of Records*—there's no such category. But that claim has been painted on the barn at the Central Wisconsin State Fairgrounds in Marshfield for years, and no one has disputed it yet!

Over 80 years ago, the Central Wisconsin Holstein Breeders Organization decided it needed a show and sale barn. The group wanted a round barn for its wind resistance, efficient stanchion arrangements and strength. Round barns used the lineal surface of lumber, which is 20 times stronger than the cross-grain surface used in traditional construction.

My uncle, Frank Felhofer, was awarded the building contract. He and his brothers began construction on Thanksgiving Day 1915. They finished the first floor, which is 150 feet in diameter, in about 6 months. The second floor was finished in 1917. It was the first and only round barn they ever built.

With no visible supporting beams, the arena is an awesome sight. It's even more wondrous when you consider that this architectural feat was accomplished without scaffolding.

Volunteer efforts have played a strong role in renovating the barn. When it needed a coat of paint in 1980, and the Fair Association couldn't afford it, I asked the second generation of Felhofers if they'd gather to paint the barn. A few of my cousins questioned my sanity, but they said yes!

Some brought their grown children and friends to help, and descendants of the original carpenters also gave us a hand. A local business donated the paint.

Today the round barn is more stately than ever. We take great pride in having it grace our fairgrounds. —Jean Wilcott Marshfield, Wisconsin

Editor's Note: This story, sent by Joan Fait of Marshfield, was excerpted from the 1995 Central Wisconsin State Fair Book. Jean Wilcott was president of the Central Wisconsin State Fair Association.

RECORD SETTER? Reportedly the world's largest, this round barn is used for showing and selling cattle at the Central Wisconsin State Fair. Stanley Fait of Marshfield, Wisconsin provided the photo.

UNUSUAL round barn was rescued by South Dakotan Jim Lacey, who moved it from the banks of the Big Sioux River to his farm west of Trent, where he restored it. Joyce Scholten, who lives nearby in Renner, provided the photo.

HANDSOME BARN graces Fred and Ferne Feikema's Triple "F" Farm near Evart, Michigan. The original owner, a railroad man, reportedly wanted a round barn because he was fascinated by his employer's roundhouse.

Our old barn, built in 1907, has weathered quite a few storms in its time, including one in the 1930's that blew off the roof. The barn looked like a cake with just the top sliced off. When the roof was rebuilt, the owners made it a little more squat.

Our barn isn't as big as some, but it's "comfortable"—60 feet in diameter and 40 feet tall. The main beams and all the upright poles and posts came from our own woods.

We've been told that the men from the surrounding area helped build it at a barn-raising. The original owner wanted a round barn because he was a railroad man and was fascinated by round-houses.

Our barn is the only round one in Osceola County. We're on Highway M66, just north of U.S. Highway 10, so the barn has been a landmark for a long time. Many people stop and take pictures of it.
—*Ferne Feikema, Evart, Michigan*

OCTAGONAL BARN above, northwest of Topeka, Kansas, has been in R. Arlen Kirkwood's family since 1913. He's made about $10,000 in improvements since 1965, including a tin roof and metal siding. "Each side is 25 feet wide and 16 feet tall. Though it likely was built in 1883 or '84, the barn remains in excellent condition," R. Arlen reports. "We hold family reunions there every year, and my daughters have had barn dances in it several times."

A DAIRY FARMER built the barn below at Red Wing, Minnesota in 1914. In 1958, Cheryl Christiansen's grandfather purchased the farm and converted the barn for use in his beekeeping business. Cheryl and husband Greg took over the farm in 1991. "We've had visitors from all over the world photograph the barn," she says. "We enjoy having such a treasure in our family." The barn is listed on the National Register of Historic Places.

ULYSSES RALPH SNIDER built this 10-sided barn for his horses near Barrackville, West Virginia in 1907. A coal company bought the structure in 1960 for use as a storage facility. A few years later, the company disassembled the barn and moved it to a new location.

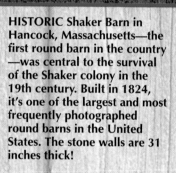

HISTORIC Shaker Barn in Hancock, Massachusetts—the first round barn in the country —was central to the survival of the Shaker colony in the 19th century. Built in 1824, it's one of the largest and most frequently photographed round barns in the United States. The stone walls are 31 inches thick!

ONCE part of a dairy farm, this barn outside Orange, Virginia was deserted as the town expanded. Now the area's senior citizens gather here to play bingo. The barn also is used as a concession stand during baseball games.

DAIRY FARMER Amos Hamilton built this circular barn on Flaggy Meadow Road in Mannington, West Virginia in 1912. Today, it houses an agricultural museum operated by the West Augusta Historical Society, which restored the structure with help from a local Future Farmers of America chapter and a 4-H club. The barn's latest improvement, a new slate roof, cost about $130,000!

W hen it comes to barns, LeRoy Schultz will drive just about anywhere for a photograph. In fact, he already has!

You see, LeRoy—whose photographs appear throughout this book—has snapped more than *15,000 barn photos* in the last 20 years! In the process, he's driven over 1 million miles, visited nearly every state in the continental United States and worn out five cars.

"I used to take pictures of every barn I saw, but after 10 years, I started getting a little more discriminating," LeRoy quips from his home in Morgantown, West Virginia.

As you can see from the photos on these two pages, LeRoy finds round barns especially intriguing. "Their rarity is the main attraction," says LeRoy, who always found comfort in the barn on the farm where he grew up. "I talked to an old man in Ohio one day who described round barns as churches. I have to agree." Amen!

FAMILY that built this one-of-a-kind barn outside Deer Park, Maryland in 1912 had a ready supply of lumber on their farm—but no money for a roof. So they topped off the structure with tar paper.

"DOUBLE" round barn near Apple River, Illinois looks just as distinctive from the air as it does at eye level. Pilots flying into Chicago and Rockford use it as a marker as they approach for landings.

"THIS octagonal barn, built in 1902 near Amherst, Ohio, is the only one of its kind in northern Ohio—and possibly the state," writes Donald Breen, president of the Amherst Historical Society. After learning in 1990 that the owners planned to demolish the barn, the society hired Amish builders to dismantle it and reconstruct it on this site. The barn will be used as a living-history museum.

"ROUND BARNS are rare in the area where I grew up, near Montevideo in west-central Minnesota," says Lois Robinson of Minneapolis. "This sturdy barn, though not in use, still stands on the farm where my aunt and uncle lived more than 60 years ago."

My family's octagonal barn, located 3 miles south of Richfield Springs, New York, has long been a landmark in our area.

My grandfather and great-grandfather built it in 1882 to provide more room for their cows and horses; they logged the pine and oak lumber on the farm.

Built with three levels, the barn followed the contour of the land in such a way that the doors on all three floors opened at ground level. Each of its eight sides is 25 feet wide, and the barn rises 60 feet from the lowest level to the top of the cupola. The cupola windows could be opened and closed for ventilation.

The upper level was for hay storage, with one chute leading to the horse stable on the middle floor, and another to the cow stable below. The main beam in the cow stable was a hand-hewn elm tree.

Although no longer used for farming, the barn still stands today and remains in our family.

—*Dorothy Ames*
Richfield Springs, New York

ROUND HORSE BARN is part of a housing development in Arlington, Washington, resident Brenda Voelker reports. The barn is owned by the development's home owners and maintained by the horse owners who use it. The barn can house 19 horses and has a tack room, stable and stall for each animal.

M y love of barns started early, with the octagonal stone barn on the farm where I grew up.

Built around 1833 near Milford, Michigan, the barn was intended to be invulnerable to fire, storms and Indian attacks, which were common at the time. Settlers' families could gather in the barn if necessary, watching for invaders through a small peephole above every door. Fortunately, our barn was never used as a fort.

My father and grandfather owned the farm from 1903 to 1960. Sometime in the early 1900's, one section of the wall fell, and Grandfather replaced the stones with wood. My father had to replace another section in the 1940's. What a thundering noise that wall made when it fell during the night!

The barn had a wooden cupola where we stored grain and seeds. The top floor held loose hay and machinery, and the basement was for the cows, hogs and grain. The barn was so big that we even had birthing rooms for several sows.

The barn was always the center of activity. In early spring, three generations would sit in the large aisle between the haylofts, singing and telling stories while we cut seed potatoes for planting. In summer, we chopped corn, filled the silo and put up hay. In fall, we gathered for threshing. It was hard work, but doing it together was fun.

My parents sold the farm in 1960. Sadly, a subdivision was later built there, and a wrecking ball claimed the barn in 1973. A road was built where the barn once stood, and its name—Stone Barn Road—keeps the barn alive in residents' memories.

—*Marti Clark, Clinton, Michigan*

RUSTIC-LOOKING round barn adds charm to farm near Wausau, Wisconsin. Ken Dequaine of Waupaca shared the photo.

RARE octagonal stone barn, shown in 1935, used to stand where aptly named Stone Barn Road now runs. Marti Clark of Clinton, Michigan shared the photo of her grandfather's barn.

F or as long as I can remember, my father, Dale Mills, has loved barns. Like other collectors, he has numerous barn figurines, paintings, prints and painted saw blades. But he doesn't stop there.

Through his reading, Dad has located *every* round and polygonal barn in Iowa, and he takes an annual "barn tour" on his motorcycle to find and photograph them. (If you own a round or multi-sided barn, don't be afraid if a man on a Suzuki pulls into your yard and asks to "shoot" your barn. It's probably my dad!)

Dad looks for barns out of state, too, during family vacations. His hobby has produced nearly four photo albums containing nothing but barn pictures. Now he's building his very own octagonal barn, something he's dreamed of for years.

We often tease Dad for "chasing barns", but I think we'd all agree his hobby has made us better people. It's helped us learn to appreciate art, architecture, history and, most importantly, quality family time. I'll never be able to pass *any* kind of barn without thinking of Dad.

—*Emily Mills, Mason City, Iowa*

A TOUR of Connie Cleaver's favorite Minnesota round barns includes (this page, clockwise from top left) the Buckentin Barn near Hamburg; the Warner Barn, built in 1880 by homesteader Peter Marsh (the barn has since burned down); a beauty owned by Steve and Judy Ellinghuysen near Altura; and the Walter O'Groske Barn near Potsdam.

When
Connie Cleaver moved to
Minnesota in 1982, she discovered
a new hobby—photographing barns.

"I just fell in love with the big old barns
there and started taking pictures," says Connie,
who now lives in Nampa, Idaho. Round barns were particularly appealing, she remembers.

"I'd never seen one before I moved there; I thought they
were fantastic," she notes. "And they're so huge! I grew up in Colorado, where the barns are small by comparison."

Connie now has several hundred barn photos; you can see some
of them in other chapters of this book. We've also devoted these two
pages to round barns she's discovered during many day-long trips
around Minnesota.

"I'd pray before I'd go, then see what came up," she recalls. "Usually,
I was pleasantly surprised—seems we'd come upon a round barn." (In
fact, Connie has photographed more rounds barns—42—than the state
historical society has listed in its records!)

"I want to leave something different for my grandchildren," Connie
says of her hobby. "I want them—and other children—to know
that these graceful old barns are out there."

MORE of Connie's favorite round barns include (clockwise from
top left) the Richard Lenneman Barn near St. Michael; the Moody
Barn, built in 1915 outside Chisago City; a circular beauty near
Crow Hassen State Park; and the St. Aubin Barn outside Watertown.

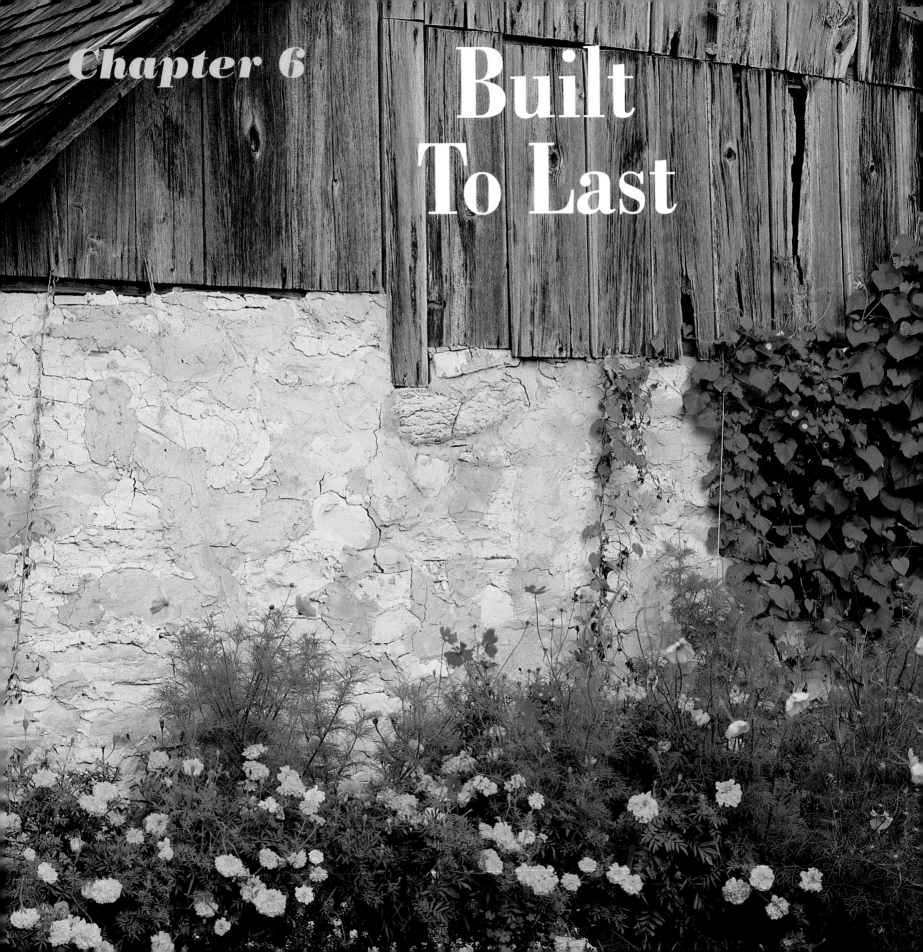

Chapter 6

Built To Last

KNOWN around Laona, Wisconsin as "Hanson's Big Red Barn", this striking structure has been a northwoods landmark for years. Tamara Kowalski of Algonquin, Illinois drove 5-1/2 hours to photograph this barn, built by her grandfather and great-grandfather in 1915.

M y father and grandfather built their barn (above) in Laona, Wisconsin in 1915, and it's been a "conversation piece" in the northwoods ever since.

Their barn was one of the first in the state without crossbeams in the hayloft, and one of the first to have windows in the basement milking parlor. In the early 1900's, most barns were kept totally dark to inhibit flies and bacteria.

The fieldstones for the basement were taken from the surrounding acreage. All the wood was tamarack, cut from virgin timber on the "back forty" and surrounding swamps. All timbers, joists, studs, rafters and posts were hand-hewn with broadaxes.

The six sets of purlin posts were erected independently. It took four teams of horses with blocks and tackles to swing them into position. The original siding is still on the barn.

To support the old-fashioned threshing machine, the threshing floor consisted of three layers of 2-inch-thick hardwood planks. The floor was so large that we could drive a team of horses into the barn, then turn around and exit the same way we'd come in.
—*Gunnar Hanson, St. Charles, Illinois*

W hen I retired, I decided to make one last visit to Grandpa's farm. He had homesteaded in Custer County, Nebraska in 1882, building a sod house and barn.

But by 1905, he had replaced them with modern wood-framed buildings. The barn became an instant landmark—and the envy of all his neighbors. I hadn't seen it since 1945, when my grandparents sold the farm and moved to town.

As we drove north on Highway 181, I had no idea what remained of the structures Grandpa had built. Then we crested a hill, and I suddenly spotted the big white barn, its eloquence untarnished by 90 weather-beaten years. It was just as I remembered it.

The current owner graciously let us roam freely around the farm site that had so enchanted us more than 50 years ago. Up close, we could see it had suffered some wear and tear, but it seemed as sturdy as when it was first built.

I think of the barn as a tribute to all the pioneers who homesteaded in Custer County. These industrious people made great sacrifices to persevere through droughts, insects, floods, blizzards and tornadoes. I'm proud to be one of their countless descendants.
—*Milan Dady, Omaha, Nebraska*

"AFTER serving in the Spanish-American War, my dad, Thomas Spickelmier, graduated from a public school in 1901, when he was 23 years old," explains Alma Knudson of Horton, Kansas. "He said he did it to learn how to figure board feet. He then cleared a heavily timbered bottomland field near Willis and began sawing the logs into timber to build the barn of his dreams (above, in 1913). As far as we knew, it was the first building he'd ever built. He simply drew up some plans and started. Rocks, sand and timber had to be hauled from about a half mile away over impossible terrain in horse-drawn wagons. Today, the barn is still in good condition."

H. GLENN MOSER built the barn at right in 1936. He's justifiably proud of the barn, which he keeps in tip-top shape. His daughter Martha says the barn and outbuildings always sport a fresh coat of red paint—even when the house doesn't!

W hen my father, H. Glenn Moser, built his sturdy barn (above) in Guilford County, North Carolina about 60 years ago, it was something of a landmark, referred to as "the big red barn on Highway 62".

Dad cut the timber off his acreage, then sawed the lumber on-site with my grandfather. Dad was an accomplished carpenter, and his handiwork must have been good advertising. After his barn was finished, neighbors asked him to build *their* barns.

Dad's workmanship has stood the test of time. And he still takes a great deal of pride in the barn; even at age 88, he still sees to it that the barn and outbuildings are kept painted—whether the house is or not!

The barn was the hub of his world as a farmer, the place where he began the day early and ended it late, milking the cow that provided milk for his six children. The barn was where he stored a summer's worth of hard work, filling the huge loft to the rafters with hay so his beef cattle would survive the winter.

—*Martha Shoffner, Julian, North Carolina*

O n a cold day in January 1902, my mother and father stood on a driveway overlooking an abandoned hop farm they hoped to buy near Oneonta, New York. Father pointed to a small hill and said, "That's where we'll build the barn."

And so a dream was born.

They bought the farm and moved there in 1903. After 5 years of hard work and planning, two men began cutting the timber. Every afternoon, Father would skid the logs to a portable steam-powered sawmill and stack them to dry.

The foundation was dug in 1909, and the two men spent the entire summer collecting *1,200 loads of stone* for it from nearby fields.

The barn builder was to start work the following spring, and he'd be on a tight schedule. The plan was to tear the old barn down and use its lumber to build the new barn. But Father didn't think

W hen my father was young, his family moved to a farm near Washington, Illinois. He was extremely shy, but a charming country lass (who later became my mother) coaxed him to join a group of young folks who walked to a small country church in Holland's Grove every Sunday morning.

In 1920, they were married in that same little Lutheran church. Several years later, the church was put up for sale. Dad's father bought it for $600, and Dad, a skilled carpenter, turned it into a barn.

Dad, who's now 98, built his own forms for the barn's foundation, borrowed a cement mixer and made his own cement, and poured the foundation with Grandpa and a neighbor.

Grandpa helped with the rafters, but Dad did everything else. He even built a cistern to catch rainwater that ran off the barn, blasting the first hole with dynamite and digging the rest by hand.

While Dad was building the barn, we were living with my grandparents several miles away down a country road. One morning my mother was busy with a new baby—me—

and mistakenly thought my 2-year-old brother, Bob, was with Grandpa. Little did she know that he'd followed his daddy to the barn.

As for Dad, he nearly fell off the barn roof when he looked down and saw Bob toddling up the lane, pulling his little red wagon!

—*Shirley Meagher*
Tucson, Arizona

WHEN Kirk Patee and Ruth Porter courted and got married in this small church (right) in Holland's Grove, Illinois, they had no idea that several years later, Kirk's dad would buy it and Kirk would help convert it into a barn (far right).

GOOD-SIZE BARN above near Oneonta, New York was built in just 6 weeks in 1910. Good thing, too—it had to be ready to store the first hay crop in mid-June. "Dad drove the first load of hay into the barn right on time," recalls Mary Sovocool of Cooperstown. A relative, Janice Downie of Oneonta, shared the photo.

the weather would permit him to turn out the cows before May 1, and the new barn had to be ready to hold hay by June 15. That left just 6 weeks to build a barn measuring 72 by 36 feet!

I still remember the builder. He looked at a diagram, then told Father where to place each pile of lumber, specifying the size, length and number of pieces. He did all this figuring in his head, despite the fact that he had *never* been to school and couldn't read or write.

At 6 a.m. on May 1, 1910, Father took a wagon to the train station to meet the builder and his 15-man crew. The crew worked from dawn to dusk building the framework, taking breaks only for meals. There were two barn-raisings, one for each floor.

The seven huge bents for the first floor were raised and secured in an afternoon. Then the workers laid the floorboards and built the superstructure of the rest of the barn.

Raising the second story was quite an event, with more than 50 men putting the massive beams in place. They worked all day, stopping to eat whenever they had a spare moment. The women had made mountains of sandwiches, doughnuts and cookies and big containers of coffee. It was like a picnic, and the men enjoyed working together.

By the end of the day, we could see what the barn would eventually look like. Then the carpenters started to put in the braces at each beam, joining and finishing the rest of the building.

June 15 was a landmark date for my father, who drove the first load of hay into the barn (above) right on time. The carpenters went on to another job while we finished haying, and they returned later that summer to finish their work. Long before the first snows, our cows and horses were at home in the new barn.

—*Mary Sovocool, Cooperstown, New York*

T he barn (below) built by my grandfather, Jacob Angst Jr., and his family in Buffalo County, Wisconsin around the time of the Civil War is a tribute to their ingenuity, energy and determination.

The spot where Jacob built his barn had a deep layer of red clay, with very hard shale underneath. The clay was suitable for making bricks, so he set up a kiln on the site and fashioned enough bricks to build a handsome three-story house.

With the clay now removed, Jacob, his wife and their nine children started chipping out the shale. It took *7 years* to excavate the site and build the barn, with the family painstakingly removing the shale, using picks, shovels and wheelbarrows. Most of the excavating was done in winter; they covered the site with straw to inhibit frost.

Locally quarried limestone for the basement walls was brought to the site in horse-drawn stone "boats". The walls were laid and mortared by hand.

The hand-hewn timbers were mortised and attached with wooden pins, forming a strong framework for the unusual driveway that runs the length of the hayloft. Sturdy wooden ramps provided access to each end of the driveway.

The barn is still in use, although mechanical elevators have eliminated the need for access ramps. —*Bergie Ritscher*
Alma, Wisconsin

BEFORE Jacob Angst Jr. and his family built this barn, they excavated a deep layer of clay and underlying shale from the site. They formed the clay into bricks, which they used to build a three-story farmhouse. Removing the underlying shale and building the barn took 7 years.

BUILDING this well-kept barn near Portland, Michigan in 1915 required a lot of tough, backbreaking manual labor, including cutting trees and moving them to a mill site and digging gravel by hand for the foundation.

When my parents decided to build a new barn on their farm near Portland, Michigan in 1915, they chose a builder known for his well-designed barns.

My brother Jerry took over the chores so Dad could start felling timber. My brother Henry and I were walking to a neighbor's farm to watch a steam-powered corn husker when we heard the first tree fall with a deafening crash. It was a large beech, and I can still identify the heavy floor timbers sawed from it.

When all the logs were cut, my sister called the builder, who lived 5 miles away. It was the first "long-distance" call our family ever made! The builder came and marked the logs, which then were skidded to a spot where a mill would be set up.

One huge elm log was moved only with great difficulty. Its size and weight can be imagined when you consider that it later was sawed into 55 rafters, each 14 feet long!

Meanwhile, Dad started digging gravel for the foundation from the face of a 50-foot sheer cliff. Undercutting the cliff was risky, as the frozen gravel above could crash down at any time. Dad was relieved when the face finally collapsed one warm February day, eliminating the danger.

When the ground thawed sufficiently, Dad and Jerry painstakingly excavated the foundation. The builder and his four carpenters arrived in April and spent a month building forms, hand-mixing concrete and pouring the foundation. They then started installing posts and beams for the first floor. By mid-June, the barn was complete enough to hold new hay.

Today, the 80-year-old barn still stands plumb and straight. It's kept in excellent condition by Jerry's son Dan and his boys, who will doubtless carry on the family's farming tradition.

Looking back, I marvel at the construction job my father undertook and the backbreaking labor performed by every person involved. These men, and millions like them, are the unsung heroes who built the United States. —submitted by Charles Leik Great Falls, Virginia; written by his late father, George

In 1929, only emergency calls were allowed after dark on our rural phone system in eastern Kansas. At about 11:30 p.m. one November night, we got just such a call, telling us that the Sweets' barn was on fire.

Augustus Sweet was one of the community's outstanding farmers. He milked 35 to 40 cows, and his farm was a showplace. My father, brother and I hopped into our Model T and took off for the farm, 1-1/2 miles down the road.

But as soon as we pulled out of our front gate, we could see the fire glowing against the starlit sky. "That barn is a goner, boys," Dad said. He was right.

Now, the Bible says that if your ox falls in the ditch on Saturday night, you pull him out on Sunday. The whole community decided the Sweets' ox was in the ditch, and we all pitched in to put them back in business.

At dawn the next morning, we heard John Deere tractors putt-putting down the road—some from as far as 5 miles away—pulling wagons loaded with straw, alfalfa, grain and cattle feed. Other farmers helped build a temporary milking barn out of straw bales, logs and brush.

Then it was time to build a permanent barn. Again, the John Deeres arrived, pulling wagons loaded with lumber, bags of cement, sand, gravel, shingles and other building materials.

Every day of the week, and especially on Sunday, everyone in the community was sawing, hammering and mixing mortar. The women made a potluck lunch every day. The whole project was like a continuing community picnic.

After church on Sundays, even the local minister joined in. "I'll be glad when this barn is finished," he noted. "I'm getting tired of preaching to empty seats."

In about 2 months, the barn was completed. We held a huge barn dance and pie social, and the new barn was turned over to a very grateful neighbor.

—*Philip Pressgrove, Lancaster, California*

My German ancestors settled in St. Clair County, Illinois in 1832, and my great-grandfather, Balthazar Knobeloch, built his barn here in 1844. The barn (at right) is still in use today and has changed little in its 152-year history.

The barn is built almost entirely of native oak, and all the beams are hand-hewn. Large wooden pins were used to connect the mortise-and-tenon joints.

The 22-inch-thick foundation walls on three sides are made of sandstone and limestone, held in place with a lime-and-sand mortar. Oxen carts hauled the rocks about 35 miles.

The original wood shingle roof lasted over 80 years until my father covered it with asphalt shingles in 1928. I put on a third layer of asphalt shingles in 1959. In 1987, I had all three roofs torn off and installed a pre-finished steel painted roof.

The farm is now operated by our youngest son, Darryl, his wife, Helene, and their three children. They live in the same farmhouse built by my great-grandfather in 1861.

My grandchildren represent the seventh generation to grow up here. Just like our sons, I'm sure they'll have many memories of playing in this wonderful barn.

—*Orville Seibert, Belleville, Illinois*

"MY FATHER, J.E. Entz, gathered stones from surrounding fields to build this barn," observes John Entz of Waterloo, Iowa. "The rocks were split by hand with a large maul. The barn was started in 1929 and completed in 1933; it was built in stages as money became available. My son, Eddie, now uses the barn to raise sheep. I think it will be here for a long, long time."

CASUAL OBSERVERS would never guess that Orville Seibert's barn has stood, virtually unchanged, for 152 years! The barn has been re-roofed three times, and one sandstone foundation wall was replaced with concrete. Otherwise, it has changed little since its construction in 1844. The farm is listed on the National Register of Historic Places.

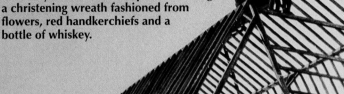

WHEN an old barn (left) on her father's Wisconsin farm was torn down, it provided some of the materials for its replacement (below), writes Carol Myers. Note the carpenter holding a christening wreath fashioned from flowers, red handkerchiefs and a bottle of whiskey.

In 1936, my father decided to tear down the old barn on his farm in Sauk County, Wisconsin and build a new one. Logs were cut from the wood lots that winter and a sawmill was set up in the pasture.

In the spring of 1937, the old barn was taken down. Some items were saved to use in the new barn, including the old weather vane—one of our most treasured possessions.

Horse teams pulling small hand scrapers excavated the building site. Masons laid foot-thick stone walls, and carpenters cut and notched the lumber.

The new barn would be 104 feet long, 36 feet wide and 70 feet tall. It would also be the first in the area to use steel window frames and be covered with galvanized steel.

About 70 men raised the barn in June 1937. The barn's "christening" was the highlight of the day. A carpenter climbed to the peak and held high a 4-foot wreath made of flowers, red handkerchiefs and a bottle of whiskey.

When the barn was finished, my parents hosted a barn dance. They bought a 24-hour dance license from the county clerk for $3—expensive for newlyweds who had just built a barn!—and scheduled the dance for 8 p.m. to 1 a.m.

More than 50 years later, we found the old license on the barn wall. Why it was still there, we don't know; perhaps it was always covered by crops. At any rate, we're preserving it so we can share the story with our grandkids. We hope they'll tell the tale to *their* families someday.
—*Carol Myers, Reedsburg, Wisconsin*

"OUR BARN was built between 1889 and 1893 by G.A. Joslin," writes Verona Floeter of Markesan, Wisconsin. "One day, one of his descendants stopped by to see where he'd grown up. He told us his father had logs for the barn floated down the Wisconsin River to Portage, then hauled here by oxen teams."

WHEN Halvor Westby (standing between horses) moved from Wisconsin to Minnesota in 1896, he wanted to let his new neighbors know about his carpentry skills. So he offered to build a barn for a neighbor, with a guarantee: If the neighbor wasn't happy, he didn't have to pay! "Grandfather was paid, and he had all the work he wanted from then on," reports Betty Hocking of Lolo, Montana. Betty's grandfather built this barn for himself in 1907.

CEMENT BLOCKS for this barn in Barber County, Kansas were made on-site. "My father shoveled all the sand into a wagon, hauled it 1-1/2 miles home, then shoveled it out again," writes Wilma Bell of Medicine Lodge. "His father designed and built the barn in 1909." The 42- by 50-foot barn remains in use today.

RUSTIC BARN and fieldstone silo outside Reed City, Michigan were built by Nicole Freeman's grandfather and great-grandfather. Nicole, of San Diego, California, says her great-grandpa made his own iron nails. Homesteaded before 1900, the farm still has hitching posts by the road.

When my parents, Bill and Julia Kenney, bought a farm in Custer County, Oklahoma in 1928, a barn came with the deal—but it was about 1-1/2 miles from the house. Bill knew that to get any use out of it, he'd have to move it closer to home.

Bill hired a mover, who jacked up the building and used horse teams to place timbers underneath. The barn then was pulled to its new site on wooden "wheels" cut from large logs.

When the government bought the farm for a dam/lake project in 1958, Bill and Julia bought a new farm 30 miles away—but it didn't have a barn. So they moved their barn *again*! This time, they had to cross the Washita River, and the bridges were too narrow to use.

Fortunately, a contractor building the dam had filled a part of the river with dirt to move construction equipment across. The contractor agreed to let the barn mover use the fill crossing.

The barn was used for another 30 years at its new location, until failing health forced Bill and Julia to sell the farm and move to town. Now all that remains of that warm, friendly old barn is wonderful memories—and the picture frames and other memorabilia made from the weathered boards salvaged by my sisters and me.
—*Betty Miller, Foss, Oklahoma*

BEN GHORLEY stands before the log barn he built with his father and brothers at Ball Ground, Georgia in 1938. "Now it's one of the few log barns left in Cherokee County," says Ben's daughter, Magdalene Tippens of Ball Ground. "Grandpa, Daddy and his brothers cleared the site by hand, cut the logs on the farm and dragged them to the site by mule. They also hand-notched each log. Daddy's now 83 and still uses the barn every day."

"ONLY HOURS after construction workers finished building our barn in 1912 near Rockville, Minnesota, a cyclone came and blew it down," recalls Ted Nieters of Greeley, Colorado. "The workers came back and salvaged what they could from the wreckage (above) and rebuilt the barn, which is still standing today (top, shown in 1933)."

I n the summer of 1966, my dad agreed to take care of my grandparents' dairy farm in northeastern Ohio while they took a long-awaited 3-week vacation.

A ferocious summer storm turned this simple task into a heart-wrenching experience. Around suppertime, we heard a violent crack of lightning that was distinctly different from the rest. We knew in our hearts it had struck something.

It had. In a few minutes, my uncle drove up and yelled to us that the dairy barn had been hit and was burning—and the phones were out.

The barn had just been filled with 1,500 bales of hay, so the fire had plenty of fuel. All the firefighters could do was try to keep the other buildings from catching fire.

When Grandpa heard the news, he asked Dad to find out if an unused barn down the road was for sale. It had been built around 1897 to store loose hay and was still in excellent condition. It had 12-inch-square beams and measured about 40 by 80 feet.

A few days after my grandparents returned, the barn was purchased and a moving company began preparing to transport it to its new home. Steel cables were rigged inside for stability, then the barn was jacked up from its foundation. I-beams with wheels attached were positioned underneath to act as axles.

The barn was too wide to move on the road, so it was moved across the fields—a feat that took a couple of weeks. This route

"OLD BARN looks like it's smiling as it moves to a renewed and useful life," writes Linda Garber. Her grandparents purchased the barn to replace one that had burned down.

124

"IN 1906, my grandfather, Richard Purdom, started to build this cement block barn," reports Paul Thomas of Bellbrook, Ohio. "He finished it in 1910—it took that long because he and a friend, Mr. Copsey, mixed all the cement using gravel from a pit on the farm. They poured the cement into a form measuring 8 by 24 inches. The barn is 40 by 60 feet, so you can figure out how many blocks it took. All the lumber—oak, hickory, ash, walnut and maple—was cut from timber that stood on the farm. Grandpa would be proud to know it's still standing." A friend of Paul's, Max Brandeberry, took the photo.

took the barn down a gently sloping hill, across a creek, then up a steady grade to the spot where the old barn had burned.

Grandpa later sold the farm, but that barn is still standing. I'll never forget the "great barn moving of 1966"! —*Linda Garber*
Westerville, Ohio

My grandfather's Pennsylvania Dutch forebay barn was built in Williams County, Ohio in 1888. Fieldstones for the foundation walls were moved into place by horses. A steam-powered sawmill was set up in the woods, and horses hauled the lumber across the frozen St. Joseph River.

The barn was raised by my grandfather, Marion Kurtz, and his friends and neighbors. Horses moved the beams into position using a block and tackle. The frame of hewn timbers was held together with oak pegs, and the siding was finished with hammered nails.

A steel roof replaced the original blue ash shingles in 1945. The Kurtz Centennial Farm is now owned by my father, age 92, and the barn remains in good condition. —*Marilyn DeVore*
Saline, Michigan

When homesteader Daniel Herboldsheimer moved to the High Plains in 1878, northeast of what is now Potter, Nebraska, there weren't any building materials readily available. But there *were* low-lying limestone cliffs 4 miles south, now known as Point of Rocks.

After casting an appraising eye southward, Daniel set out for the breaks and began quarrying the stones with a crowbar, pick, shovel and log chains. Day after day, he chiseled the stones into shape and hauled them home.

In 1888, he had enough to build a stone milk house. A few years later, he replaced his sod house with one of hewn logs and stone.

In 1898, he excavated an 8-foot-deep reservoir and lined it with stone, then completed a stone barn that was 66 feet long and 32 feet wide. His pride and joy was the stone corral, which is 2 feet thick, 8 feet tall and 120 feet square.

In an interview in 1942, Daniel said, "I wouldn't trade my place for any in the country—in the world, for that matter. I've got 13 miles of fence with not a boughten post on it. I've got stone buildings that are warm in the winter and cool in summer and will last for centuries to come.

"Sometimes I look at the cliffs over yonder and wonder why more farmers don't use some of that stone for their buildings. They wouldn't be sorry, I can tell them that."

The barn, corral and six other stone buildings—all located near our farm—are now listed on the National Register of Historic Places. The barn is still used for livestock.
—*Ruth Morgan, Gurley, Nebraska*

STONE quarried from low-lying cliffs 4 miles away provided building material for Daniel Herboldsheimer on his ranch near Potter, Nebraska. In the 1890's, he built a 66-foot-long barn and a 120- by 120-foot corral whose 8-foot-high walls are 2 feet thick!

Barn Raisings...

When the large barn at well-known Malabar Farm State Park burned down in April 1993, our family—as well as many others in our area of north-central Ohio—was heartbroken.

We've lived nearby all our lives, so we've always taken a keen interest in the famous farm, once the home of the late Pulitzer Prize-winning author Louis Bromfield. We've attended many of the special events held there each year.

The 103-year-old barn was a majestic old structure, and we were delighted to hear that the Timber Framers Guild of North America would build a near-replica of it on Labor Day weekend 1994.

On that weekend, we sat for hours in a large grass pasture nearby, watching members of the guild raise the barn the way it would have been done 100 years ago (see photos at left).

We watched intently as the posts and beams were joined with handmade wooden pegs. Instead of the whine of electric power tools, the predominant sound that day was the pounding of wooden mallets.

One interesting note—the main beam in the front of the barn has a lightning bolt carved in it, a good luck charm against lightning or fire.

After the second day of work, folks celebrated the occasion at a barn dance held under the wooden framework and a star-filled sky.

We feel very fortunate that our family witnessed this event; I doubt we'll ever witness another barn raising. And every time we see that barn, we'll remember how the great barns of old were built by hand.   —*Starla Scholler*
Butler, Ohio

MEMBERS of the Timber Framers Guild of North America raised a barn in September 1994 at Malabar Farm State Park near Mansfield in Richland County, Ohio. Many hands make the task look deceptively easy, as you can see in this series of photos snapped by Starla Scholler.

126

I arrived at Vernon Kline's farm near Shreve, Ohio at 7:30 a.m. on June 4, 1994 to watch an Amish barn raising. Though it was early, one wall section, or bent, was already erected, the barn floor was swarming with workers and Amish buggies were parked across the road.

By 9:30, the frame was up, exterior siding was being attached and rafters were being installed. The ladies watched and socialized from the lawn before preparing dinner.

Small boys in straw hats carried scrap wood away, older boys started nails in siding and older men built doors in the shade of trees. Before noon, the steel roof was being attached.

I was introduced to Vernon's brother, David, who invited us to visit his farm. We did so, and later, when we returned to Vernon's farm, the barn stood as though it had been there for years.　　—*Charles Leik, Great Falls, Virginia*

...Are Stirring Sights

Recipe for a Barn Raising

Martha Lebo of Elkton, Maryland shared the following excerpt from Mary Emma Showalter's Mennonite Community Cookbook, *Favorite Family Recipes*: "This bit of information was found in a quaint old handwritten recipe book from Great-Grandmother's day. A barn raising was quite an event during those early years. When a new barn was built, all friends and neighbors came to help.

"Homemakers of our day will no doubt be astounded at all the food consumed in 1 day. It's even more difficult to believe when you consider that most of it was made in Great-Grandmother's kitchen! Here is the list:

115 lemon pies
500 fat cakes (doughnuts)
15 large cakes
3 gallons applesauce
3 gallons rice pudding
3 gallons cornstarch pudding
16 chickens
3 hams
50 pounds roast beef
300 light rolls
16 loaves bread
Pickled red beets and pickled eggs
Cucumber pickles
6 pounds dried prunes, stewed
1 large crock stewed raisins
5-gallon stone jar white potatoes and the same
** amount of sweet potatoes**

"This made enough food for 175 men."

MASTER BARNWRIGHT Josie Miller (in foreground of top photo, square in hand) and his crew from Holmes County, Ohio have raised over 500 barns, including the one shown in this series of photos. Charles Leik took the photos near Shreve, Ohio in June 1994.

Chapter 7

Century-Old Sentinels

y husband and his brother own a barn (right) that's been in their family for eight generations—and was built between 1683 and 1690!

Captain Ephraim Flint built the 31- by 34-foot barn after inheriting 750 acres in Lincoln, Massachusetts from his father. Like many farmers of the time, Ephraim built his barn first, then constructed a saltbox house.

One of his descendants added another section to the barn between 1750 and 1760, bringing its length to 68 feet.

In 1918, the barn was moved across the road. My husband remembers watching as the barn was raised and placed on large rolling logs. It took two horses, a capstan and several men to move it to its current location behind the old homestead.

The old barn withstood Hurricane Cora in 1938 and Hurricane Dora a few years later.

Inside the barn is an interesting historical artifact. In 1940, Henry Flint won three plowing contests sponsored by the Ford Motor Company. Henry Ford himself presented my brother-in-law with his prize—a new Ford tractor, cultivator and plow. The tractor is kept in the barn and is still used on the homestead.

—Gwen Flint, Claremont, California

HISTORICAL BARN above in Lincoln, Massachusetts was built between 1683 and 1690. It's been in the Flint family for eight generations.

OVER 200 years old and still going strong, this barn, owned by John Budd Jr., stands near Kintnersville, Pennsylvania. John says it was built around 1775. The red shale structure was used continuously as a barn until about 1935. The round "portholes" provided ventilation. John knows of only one other barn in the area with such windows.

BUILT in 1885, barn at left outside Flint, Michigan was rebuilt in 1983 as part of an old-fashioned Amish barn raising, notes Sue Kean of Innerkip, Ontario.

MAIN SECTION of barn at left has stood unchanged since 1862, when the bloody Civil War battle of Antietam was fought right outside its doors. R. Earl Roulette of Sharpsburg, Maryland says his great-grandfather's family hid in the basement of their farmhouse during the battle. When the fighting ended, the barn was used as a hospital. Some 700 soldiers from both sides were buried on the farm, now part of the Antietam National Battlefield.

BARN at right in Montana's Smith River Valley was built around 1885 and remains fairly solid, even though the lack of a foundation has caused some settling and decay. Jim Fuller of White Sulphur Springs says the original board-and-batten roof lasted 100 years before it had to be replaced. The barn is on what was once Jim's grandfather's homestead.

FRONT SECTION of Eldon Olsen's barn near Pine Island, Minnesota was built in 1870 on shale rock, with the siding running lengthwise. The rear section, added in 1942, was made of concrete blocks and vertical batten-board siding. "We believe the addition was done this way because of the shortage of building materials during World War II," says Eldon's sister, Mrs. John Strom of Westminster, California. The structure is 90 feet long and 32 feet wide.

IN 1877, Charles Bronson built this large barn of native limestone in Osage County, Kansas. It's 80 feet long, 50 feet wide and 25 feet tall, and is believed to have been built to house circus animals during the winter. The owners, Mr. and Mrs. Wilson Karnes of Overbrook, have lived on this homestead for 60 years.

"MY GREAT-GRANDFATHER built this barn in Deerfield, Massachusetts in the early 1800's, and it's still in use," writes Ruth Herzig of South Deerfield. "The farm animals and crops are long gone, but the seventh generation of our family is making its own happy memories here, just like those before them."

A CIVIL WAR COLONEL built this barn after retiring and buying a farm near Pine Grove, West Virginia, says LeRoy Schultz of Morgantown. For 20 years, the barn's first floor housed the community's first post office.

In 1934, my father and I needed a better place for our horses on our Ohio farm, and figured we could build a lean-to on our corncrib for $100. We took our plan to Bert Dixon, who had built several barns in the neighborhood.

But Bert said that for $100, we could buy an *entire barn* that was sitting along the railroad tracks in Radnor, Ohio. The barn had been used for the Chicago World's Fair in 1893-94. After the fair, it was moved to Radnor and reassembled along the tracks to store hay that was loaded onto railcars.

To make a long story short, we bought the barn (at right) and had it hauled on a truck to our farm near Powell. In 1936, Bert reframed it, raising the floor for a hayloft and leaving the lower space for cattle. —*Craig Askins*
Powell, Ohio

FOR $100, Craig Askins and his father were ready to build a lean-to on their corncrib—until they learned they could buy a whole barn for that amount! It's now over 100 years old.

BELOVED "Dr. Pierce Barn", now owned by Roger Norr, is photographed almost every day. "There are paintings and prints of it everywhere, particularly in doctors' offices," Roger says. "When we bought the property in 1992, neighbors and strangers alike warned us *not* to tear it down!" Photo was taken by Scott Smith of Logan, Utah.

STONE BEAUTY built in 1820 is still standing proud outside Unionville in Chester County, Pennsylvania. The stalls inside were lined with copper so the horses couldn't chew the wood, says Jerry Irwin of Paradise, who shared the photo.

When people think of Utah's Cache Valley, they may not associate it with the huge "Dr. Pierce Barn" (above) along Highway 89-91 in College Ward.

But to local residents, it's a beloved landmark—and one of the most-photographed barns in the area. My grandfather, Lovenus Olsen, built the barn in 1904.

Dr. Pierce's representatives spotted it in the 1920's and thought that its size and high visibility would make it the perfect spot for one of their ads. In exchange for letting them paint his barn with the ad, they offered Grandpa $10 a year, plus an annual sign re-painting. The ad has been there ever since.

The tonic hasn't been available for some time, and the message hasn't been repainted in years, but the barn is still there.

In those early years, local residents frequently referred to the barn in giving directions to travelers. In a rural farming area with no numbered addresses, the barn was a foolproof guidepost.

But the barn was much more than just a billboard or signpost. It was a vital part of Grandpa's farming operation, helping him support his wife and 10 children.

When the property was sold, the new owners thought about tearing down the rickety structure. But friends and neighbors talked them out of it. It's even more weathered and worn today. But for now, the citizens of Cache Valley can still point with pride to "their" barn. —*Luann Pehrson, Millville, Utah*

GOLD mined in California enabled an Iowa farmer to build this handsome bank barn in 1866. The farm is now owned by Tjada Sweers, whose family bought the land in 1903.

The "Gold Rush Barn" (above) has been in our family since 1903, but its colorful history began decades earlier.

As the story goes, a penniless man named Irving Whitney went to California during the Gold Rush, hoping to find enough gold to buy a farm. Each night when he returned to his boarding-house, he tossed his gold-filled saddlebags into a corner. He figured that if he never took the bags to his room, no one would suspect they contained anything valuable.

Mr. Whitney's plan succeeded—he returned to Alden, Iowa, ex-

"OVER 100 years old, our barn is one of the oldest in the community," observes Helen Whittaker of La Rue, Ohio. Since the Whittakers purchased the farm in 1959, family members and friends have replaced the barn's foundation, given it a paint job and restored the cupolas.

UNUSUAL CUPOLAS decorate this barn near Accident, Maryland. It was built in the 1850's. LeRoy Schultz of Morgantown, West Virginia says it is now used as a service station.

EVEN AS A CHILD, De Loris Groene of Winfield, Kansas admired this Chautauqua County stone barn, which sat across the road from her family's farm. The barn and a matching house were built by Ambrose Yancey, probably in the late 1800's. Matthew Fretz of Hope, Michigan provided the photo.

changed his gold for cash, bought a farm and built his barn.

The basement's 18-inch-thick walls are made of flat stones taken from a riverbed about 2 miles away.

—*John Sweers*
Alden, Iowa

Around 1881, community leader Jonathan Robinson built "the largest barn around"—a three-story structure 80 feet long and 60 feet wide—on a farm his ancestors had received in a land grant from William Penn in the 18th century.

The hand-hewn oak logs used to frame the interior are 8 inches square, held together with wooden pins. The exterior siding of yellow pine has weathered into beautiful shades of gray.

The barn is sturdily built and has withstood many a storm—a real trib-ute to its builder. Except for necessary repairs, like replacing the original wood-shingle roof with tin, little has been changed.

Today, my husband and I use the barn to store hay, feed cattle and run our Christmas tree operation. Whenever I go there, feelings of nostalgia sweep over me. The sweet-smelling hay, flitting pigeons, antique grain cleaners and occasional barn owl remind me of my childhood on the farm.

—*Myrtle Haldeman*
Clearville, Pennsylvania

STURDY Pennsylvania barn, built around 1881, houses hay, cattle, and Dan and Myrtle Haldeman's Christmas tree business. It's located on Big Creek Road just north of Purcell in Bedford County.

The barn (at left) on the Rhode Island farm where I grew up was built in the 1840's. It's held together with wooden pegs and has three levels—a cellar, the main floor and a haymow.

Now it houses antique buggies and cars, including my father's 1922 Pierce Arrow touring car. I've added a lounge area to the barn, where we gather with neighbors to listen to old records, eat and tell stories. The old swing in the barn door still gives a good ride, just as it did when I was a kid.

I love this barn not only because of the memories of playing there, but because I feel peaceful in it. Just looking at it gives me a wonderful sense of calm.

In the near future, I plan to smooth the main floor and have a neighborhood dance and family reunion. This barn is going to remain vital and alive as long as I can manage it.

—*Gardner Rogers, Little Compton, Rhode Island*

The barn of my youth was a huge three-story stone structure dating back to 1727. It was the fifth and largest barn built on the Durham Plantation in Bucks County, Pennsylvania, and its builders were master craftsmen. Known simply as Barn No.

GARDNER ROGERS has converted one area of his family's historic barn into a lounge for visiting with neighbors. Part of the barn still holds the antique tools used by his grandfather, who purchased the farm in 1893, and his father, who was born there 2 years later.

BATTERSON BARN is a beloved landmark to travelers on County Road 74E between Livermore and Red Feather Lakes, Colorado. Built in the 1870's, it has been well-maintained over the years. Mary Dubler of Livermore provided the photo.

LIMESTONE BARN about 2 miles south of Filley, Nebraska is believed to be the largest of its kind in the state. Elijah Filley built it for $3,500 in 1874. The stone came from Elijah's own quarry on nearby Rock Creek. The stone walls on the 54- by 44-foot barn are 2 feet thick. It's listed on the National Register of Historic Places and is owned by the Gage County Historical Society. Board member Edvert Aden of Wymore sent the photo.

5, it has endured for more than two and a half centuries!

At the time the barn was built, the Colonists were being taxed heavily by the British. That's why the barn measures 40 feet by 98 feet. Any building with a foundation longer than 100 feet was subject to additional taxes, so the builder cleverly kept it short of that mark.

Three sides of the barn are made of fieldstone and iron ore, and the walls are 2 feet to 3 feet thick. Gigantic hardwoods felled in surrounding forests were hand-hewed into beams and rafters, which still bear visible impressions from the tools. Wooden pegs hold the rafters together.

My grandfather, Harvey Riegel Sr., bought the farm in the early 1920's. When my parents married in 1926, they made the farm their homestead.

Today, the beautiful old barn stands silent, the lowing of animals audible only in memory. There is no clattering machinery, no bustling farmer, no stored grasses and grain. The barn was bought by people from the city, and they're turning all three floors into apartments.

But the barn is still there. From distant roads, we can gaze across the fields at her, admiring her beauty and charm.

—*Lorretta Deysher, Bloomsburg, Pennsylvania*

"MY GREAT-GRANDPARENTS and grandparents built this Pennsylvania-type barn in 1882 near Cedar Falls, Iowa," observes Virginia Shultz of Pottstown, Pennsylvania. "It was built from stone quarried on the farm."

HISTORIC limestone barn near Fredericksburg, Texas has been converted into an inviting gasthaus filled with antiques and folk art by Bill and Leigh Waller. Their Palo Alto Creek Farm was homesteaded by German immigrant Karl Itz in 1875. The barn, built around 1880, is just one of the original buildings remaining on the property.

IVY-COVERED fieldstone barn on Highway 57 outside Cedar Falls, Iowa is listed on the National Register of Historic Places, says Rosie Peterson, executive director of the Cedar Falls Historical Society. William and Charles Fields built the barn in the late 1800's for their prosperous stock-breeding farm, which sprawled over 3,000 acres and employed 30 people. The barn is now owned by Marshall Uhl, whose family purchased the farm in 1906.

SETTLER BARN at left, built by pioneer Nicholas Eckoff in 1859, is one of the oldest barns still standing in Menomonee Falls, Wisconsin, say owners Jim and Bonnie Ziolecki. While restoring the barn, Jim had old telephone poles cut to size on antique equipment being used for demonstrations at a local vintage tractor show. This gave him the extra-long boards he needed, as well as "period" saw marks.

STONE BARN on Gary and Michele Achey's farm near Hellertown, Pennsylvania dates to 1849. "It's remarkably intact inside and out, with massive exposed stone walls and large hand-hewn beams," Michele relates. "Some of the old pulleys and ropes for hoisting hay are still in place, too."

PIONEER cattle baron Charles Goodnight built this barn on his sprawling ranch west of Pueblo, Colorado in 1870. The limestone rocks were quarried from nearby canyon walls, and the rafters and beams were made of hand-hewn native timber. The barn is listed on the National Register of Historic Places, and efforts are under way to restore it, says Arla Aschermann of Pueblo, who shared the photo.

BUILT FLAT on the ground in 1854, barn at left near Leeper, Pennsylvania was raised and converted to a bank barn with a stone foundation about 1891. Current owners John and Lois Callahan rebuilt the stone and earth ramp in 1992, using huge stones from another barn foundation.

BUILT IN 1889 by Amish farmer Abner Miller, who moved here from Kansas, this big bank barn stands about 4 miles east of Hubbard, Oregon, reports Charity Laib of Salem. "Abner sold his farm to my grandparents, Amos and Delilah Troyer. When I was 8 years old, my family moved there because my grandparents couldn't manage anymore. The barn is a community landmark."

"AFTER YEARS of research, I traced my family tree back to my Great-Great-Grandfather Leeson's homestead in Varney, Ontario, where this barn stands," writes Laura Billings of Niagara Falls, New York. "It was built in the 1860's. I brought home a rock from the foundation and placed it in my backyard pond."

The limestone walls of our 100- by 40-foot barn have survived a tornado, a fire and 115 years of use!

Thomas Ferguson, a Scottish stonemason and farmer, moved to his homestead west of Chatfield, Minnesota in 1871. His family lived in a log house while he laid every stone for a two-story house and this monumental three-story barn. The barn was built in two sections, with the last section completed in 1912.

Today, visitors enjoy driving through the archway in the middle of the barn.

—Deb and Gary Anderson
Chatfield, Minnesota

A SCOTTISH MASON built this spectacular bank barn from Minnesota limestone in two stages, completing the last section in 1912.

BEFORE Wolfsville, Maryland had a church building, it had the Hoover Barn, where services were held from 1819 to 1847. The stone walls are 2 feet thick.

The over 175-year-old stone structure on our family's homestead near Wolfsville, Maryland is more than just a barn. For nearly 30 years, it also doubled as a house of worship.

When the barn was built in 1819, Wolfsville had no church buildings, so services were held in the Hoover Barn. This continued until 1847, when the present United Brethren Church was built.

Church services for the barn's centennial celebration in 1919 drew the largest crowd ever assembled in Wolfsville, and subsequent anniversary celebrations drew large crowds, too. The congregation's original Bible and pulpit were used each time.

My sister still lives on the farm, the barn still stands and the Brethren congregation observes the anniversary of the barn's construction every year.
—Elizabeth Biser
Pittsburgh, Pennsylvania

FAMED UNDERGROUND RAILROAD used this barn 5 miles southeast of Mendota, Illinois as a stop, says Loretta Dewey of Mendota. The barn was constructed in 1838.

CODDINGTON BARN outside Glenwood, Iowa is part of the Mills County Historical Museum, housing antique farm implements—and even a life-size, hand-carved wooden mule! Built in 1875, it was dismantled, moved to its current site and reassembled in 1988. The windmill also is from the original farmstead, according to museum board member Lester Hunt.

DONNA TYINK doesn't know exactly when her barn was built, but she knows it survived a fire that destroyed much of Holland, Michigan in 1871. "Barns like this are fast disappearing in our area," Donna laments. "Many that were landmarks when I was growing up are now gone or being torn down. They're such an important part of our history."

STONE BARN was built by John Dickenson and his sons after homesteading near Robinson, Kansas in the early 1850's. The 3-foot-thick walls are made of river rock hauled by wagon from Atchison. When the farm was sold in 1979, local residents feared the historic barn might be razed, but the new owners—Dennis and Stanley Tietjens—promptly restored the structure, says Selmer Anderson of Robinson. "They even put on a new roof after a windstorm took half of it off," writes Selmer, who shared the photo.

STURDY BRICK BARN was built by Z.T. Dunham, son of the first white settler in Crawford County, Iowa, near Dunlap in 1870. It's now owned by the Dunlap Historical Society, which plans to use it to offer displays of blacksmithing, weaving, soap-making and other frontier crafts and trades. The barn is listed on the National Register of Historic Places, says Doris Bingham, society secretary.

"OUR LOG BARN was built by my grandfather, Lewis Fourt, around 1880 on land my great-grandfather homesteaded about 1830," writes Mildred Melton of Houston, Missouri. "This land has never been owned by anyone other than Fourts."

DORMER of barn built in 1881 by Andrew Rieder near Delano, Minnesota was used as a pigeon house. In later years, the dormer was a popular play area for several generations of Rieder children. Current owners Dan and Dave Rieder have turned the barn into an attractive woodworking shop, reports Elly Bruhn of Delano.

Chapter 8
Fun and Festivities

WHILE DATING, Dale and Lillian Peckham (at left) danced in the barn behind them near Ypsilanti, North Dakota. They later held their wedding reception there in 1944 and celebrated their golden anniversary there in 1994 (above). The Peckhams received a national award for barn preservation in 1992. Son LaRell now uses the barn for calving and storing hay and straw.

When my parents, Dale and Lillian Peckham, were dating, they attended dances that Grandpa hosted every Tuesday in the hayloft of his barn (that's Dale and Lillian and the barn above). When they married, they had their wedding dance in the hayloft, too. So it seemed only natural to celebrate their 50th anniversary there!

When the barn was finished in 1938, the neighbors urged Grandpa to have a dance. He did, and it went so well that he held dances every Tuesday, May through October, for the next 10 years.

Admission was $1 for men and 50¢ for women. Big Bands performed about five times a summer, and local musicians provided music the other nights.

Each dance drew 200 to 300 people, some from as far as 60 miles away. On July 4, 1946, after World War II ended, *1,400 people* packed the place to the rafters!

As our parents' 50th anniversary approached, my siblings and I decided to have the celebration in the hayloft, where their married life began. We started getting together every weekend to clean the floor and prepare for the party.

Our parents said we were working too hard, and it would be easier to just rent a place. But when the big day arrived and we saw how much Mom and Dad were enjoying themselves with nearly 200 relatives, friends and neighbors, we knew it was worth the effort.

We set up tables and chairs on the dance floor, provided plenty of food and played Big Band music, waltzes and polkas. Many of the guests had attended dances at the barn in their younger days, so there was lots of reminiscing. One neighbor said that of all the golden anniversary celebrations she's attended, the one in the Peckham Barn was the best!

—Charlene Messer
Hannaford, North Dakota

EVERY DECEMBER, Mr. and Mrs. John Larson host a live Nativity scene and worship service in their barn near Hollandale, Wisconsin. "We have sheep, calves, Joseph and Mary, and a real baby portraying Baby Jesus," Mrs. Larson explains. "We sing Christmas songs, and our pastor talks about how much this simple barn was like the stable in Bethlehem." After the service, friends and neighbors arrive to share cookies and hot cider. The Larsons' son, Randy, took this photo.

"MY SON GARY married his wife, Andrea, on September 1, 1991 in the barn on our 80-acre farm southwest of Plymouth, Wisconsin," relates Ruth Jaeger. "We had to clear out 30 years' worth of 'stuff', put on a new roof and replace some timbers, boards and the barn doors. All the guests agreed it was a very unique wedding—right up to the mouse that ran up the wall during the reception!" Dan Hebecker of Plymouth took the photo.

My grandfather, Ray Ward, built his barn in Crystal Springs, West Virginia around 1906. It's the oldest one in Crystal Springs—a community named for the waters my great-grandfather discovered when he settled there.

Grandfather was a dairyman all his life, and had the largest dairy in Randolph County. He had 10 children, and his barn was a focal point for the family—and still is.

My aunt and cousin still live and work on the farm, so we migrate back to it from all over the country for family reunions. We camp in front of it or sleep in the hayloft.

The kids swing from the big rope in the loft and drop into the yard as they play tag, "pirates" or whatever games their imaginations dictate. We tell stories, reminisce and sing around a campfire in front of the barn.

To the entire Ward family—over 100 of us—the old barn represents our roots, a sense of permanence in a quickly changing world and our family heritage.

—Connie Garnett
Elkins, West Virginia

REUNIONS for Connie Garnett's family always center on the 90-year-old barn her grandfather built around 1906. The barn got a fresh coat of paint shortly after this photo was taken.

Barns are for crops and animals, of course. But when I was a child, our barn was for *me*. On a rainy day, it was a wonderful retreat where my imagination could run wild.

When I was 8, the barn posts made for great cover as I tried to capture "bad guys" without getting "shot" myself. My first "tree house" was constructed in the barn's rafters. I even rigged up a "telephone" of tin cans and wire. There was one problem—since I was an only child living in the country, there was never anyone to talk to on the other can!

The barn continued to be a refuge for me as I got older. It kept me dry and provided a place where I could be alone to think. I had privacy there like nowhere else.

In 1963, when I needed a place to practice my lines for the senior class play, I went to the barn, which was filled with the summer's harvest of burley tobacco. There, surrounded by that unique smell, I learned my lines for *I Remember Mama*. Somehow, studying remains one of the most significant things I ever did in the barn…my barn.

Today, I'm still too far away to enjoy the barn's solitude and privacy, or the aroma of tobacco curing in autumn. The sound of rain on the metal roof during a summer shower is miles away. But the barn is still there.

The barn was built to cure tobacco. But within those grayed oak walls, I was the one who was cured the most.   —*Dennis Burrows*
Lexington, Kentucky

When I was growing up in northern Illinois, a family from our church hosted a Halloween party in their barn every year.

Dressed in costumes, we feasted on hot dogs, chips, apple cider and doughnuts, then piled onto wagons heaped high with hay for a ride over country roads. We sang until we were hoarse and sometimes we nearly froze, but we didn't care. We just burrowed under the hay to keep warm.

When we'd return, the barn had been transformed into a "fun fair" with game booths, a haunted house, a maze made of hay bales and the "death chute"—a slide set up in a hay window. The ride down was terrifying but fun, and we landed safely in a big pile of hay.

The big finale was a scary movie. We'd sit on one side of the hayloft while the movie was projected across from us on a big white sheet. Even the projector breakdowns added to the excitement, with the barn plunged into pitch darkness and everyone screaming and making ghostly noises.

I'm sure that barn was quite average, but at Halloween, it became a wonderful place, full of excitement for a young girl who lived in the suburbs but was "country" at heart. Those nights gave us memories to talk about for years.   —*Cheryl Carter*
Colorado Springs, Colorado

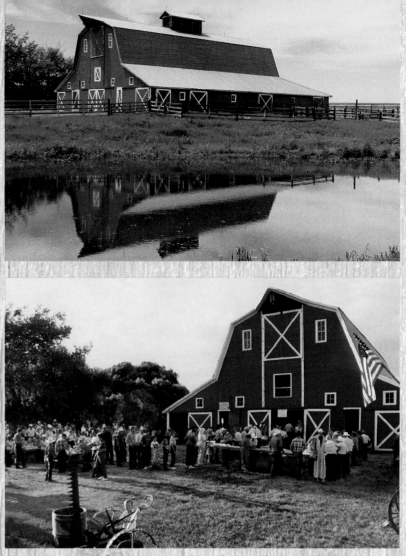

"THE BIG RED BARN" at top owned by Gloria and Raymond Beedy was built on the Karl Martin ranch near Genoa in Lincoln County, Colorado in 1916. The Martins hosted barn dances regularly, and Gloria is continuing her family's tradition. She and Raymond host an annual barbecue and dance (shown above and below) to benefit local hospitals and nursing homes, and deck the barn with festive lights at Christmas.

To Dad, our barn (right) north of Bucyrus, Ohio was a place to milk cows and store livestock food and machinery. But to my five brothers and me, it was a giant playhouse.

One of our favorite activities was to build a "bale house". My brothers would lay out a maze of tunnels with bales of hay and straw. We'd then crawl through the tunnels on hands and knees.

Since it was dark inside, we never knew when we'd run head-on into a dead end, or fall several feet into a drop-off. The scarier it was, the better!

Being the youngest as well as the only girl, I always hesitated to enter a dark opening. But my sense of adventure always helped me overcome my fear.

My dad often scolded my brothers for messing up the haymow, but he put up with the inconvenience during our growing-up years. He, too, knew that there was no more exciting place to play than in a barn.

My dad has now sold the farm to his grandson, who has a new baby. Looks like someday there'll be another crop of kids playing in the barn. —*Dorothy Hsu, Columbus, Ohio*

HAYMOW in this barn near Bucyrus, Ohio was a giant playhouse for Dorothy Hsu and her five brothers. One of their favorite activities was making scary mazes of tunnels out of bales of hay and straw.

When I was younger, I'd help put load after load of baled hay into our barn in Wisconsin. Sometimes we'd stack as many as *13,000 bales* right up to the ceiling!

With the bales piled that high, it was easy to climb into the cupola and marvel at the lofty view. The hired hands and I would think nothing of leaping from dizzyingly high stacks of bales onto mounds of loose hay, or of performing daring balancing acts atop the oak cross beams high above.

Sometimes, at the end of the day, we'd have hay bale-throwing fights. These were much like snowball fights, except we'd earnestly test our strength by hurling bales of hay. Those were some times. —*Paul Hanson*
Whitehall, Wisconsin

BARN near Whitehall, Wisconsin provided the setting for hours of fun when Paul Hanson was growing up—including hay bale-throwing contests!

FOLKS who live around Windom, Minnesota know Bernard Soleta's barn pretty well. Years ago, Lawrence Welk and his band performed at several dances there! Edith Buhler of Windom provided the photo.

A boyhood friend lived on a farm with the most up-to-date barn I remember. It had a spacious driveway, stalls for eight horses, storage bins, a tall silo, running water, roomy calf pens and a huge hayloft.

It also had something I'd never seen on another barn—a huge ventilator atop the rounded roof. The ventilator was about 6 feet square, with angled slats to let air circulate while keeping the rain out.

It provided a cool, airy room with ample space for meetings of my friends' secret "Sky-High Club". It was a long, hard climb to the club room, but getting down was easy. We just dropped into a heap of soft hay.

Today, barns are going, going...almost gone. But they've left many of us with warm and happy memories. —Frank Bennett
Garnett, Kansas

A s young girls, we often played "pocketbook" in a neighbor's dairy barn. We'd tie a long string to a purse, place it by the roadside and hide behind the closed barn doors, holding the other end of the string. If a motorist stopped to pick up the purse, we'd pull it back toward the barn.

Most people took this joke good-naturedly, but one man became very agitated and started to walk toward the barn doors when he realized what was going on. We were scared half to death! As he got closer, his ire receded, and he returned to his car and drove away. We played very little "pocketbook" after that.
—Loise Fritz, Lehighton, Pennsylvania

I have many happy memories of playing hide-and-seek in our barn at milking time, before electricity lit the dark corners and milking machines broke the quiet. There were 10 children in my family; the older ones helped milk while we younger ones played.

I remember the barn's serenity, and its pleasant sounds and scents—contented cows chewing and stanchions creaking...the sweet-smelling hay...the flickering, dancing lantern light.

The family farm was a wonderful place to grow up. There's something very special about a family working together, and my mom was always there. —Loretta Troyer, Mio, Michigan

O ur family's barn (below) near Delphi, Indiana was a great place to play barn tag.

For those of you who don't have five brothers and a lot of cousins, here's how you play: First, find an old soccer ball. Then pick one person to be "it".

Everyone then hides in the barn, and the one who's "it" tries to find them. When he or she finds someone, they must hit that person with the ball.

This is a rather tame description of a wild game. "Hiders" have been known to leap out of the hayloft to keep from getting "tagged". Once a hider almost rolled off the barn roof!

As a mother, I often tried to discourage the kids from playing, so if they got hurt, they'd never admit it was from barn tag. Scrapes and bruises were always explained away, except once when a big red ball imprint showed clearly on a boy's face!

The boys are grown now, but when they come back to the farm, their fondest memories are of playing barn tag. —Michaelle Wood
Delphi, Indiana

ROUSING GAMES of barn tag kept the walls of this barn in Indiana ringing with the sound of children shouting and laughing.

I n 1888, Wilbur Emery Campbell hosted the biggest event Kiowa, Kansas had ever seen—in the biggest barn anyone had ever seen.

Mr. Campbell was a prosperous rancher and racehorse breeder who at one time owned 489,000 acres in south-central Kansas and the Oklahoma Territory. In 1888, he moved his family to Kiowa and built a two-story house and an enormous barn.

The main section was 100 feet long and 40 feet wide. A three-story extension on the west end measured 40 feet square, and an extension on the east was 200 feet long and 245 feet wide.

On Thanksgiving Day, the Campbells hosted what the local newspaper described as a "joyous carnival and ball, fit for a queen's entertainment" to dedicate their new home. The Campbells sent out 5,000 invitations!

Heaters warmed the barn's first floor, where the banquet was served. Chairs were brought from the Kiowa Opera House. Portraits of the ranch's celebrated horses and cattle hung on the side walls of the dining room, with large photographs of the Campbells at one end.

"The ball and banquet surpassed anything of the kind ever held in Kiowa," the newspaper reported. "The entire upstairs of the huge structure was brilliantly illuminated, and all who wished could dance. Several times, more than 100 persons were on the dance floor at once. The music was excellent, the supper beyond description—grand."

Mr. Campbell retired from ranching in 1903, after selling off all his land in parcels. In later years, the deteriorating barn became impractical for modern use. It was torn down with regret by the present landowner, Huber Farney.   —*Jean Brown, Kiowa, Kansas*

"THE HILL BARN has been in our family for many years, and has been the site of many celebrations," says Christina Bechter of Mount Bethel, Pennsylvania. "The dances held there years ago (above) were famous, and we've had many family reunions and parties there. The largest celebration was my wedding (top photo) in 1990."

IN 1888, this barn hosted the biggest event folks in Kiowa, Kansas had ever seen. To dedicate his new home, wealthy rancher and racehorse breeder Wilbur Campbell invited 5,000 folks to a lavish ball held in the huge barn. Photo from Mildred Farney of Kiowa.

IMPRESSIVE RED BARN, built in 1948, is nestled under tall California redwoods in Crescent City and overlooks the Pacific Ocean. Its hayloft was converted to a basketball court over 30 years ago.

Our barn has housed cows, horses, goats, rabbits, chickens and a variety of other farm animals. But in the summer of 1995, it housed our grandchildren during a family reunion.

We transformed the loft, which was converted into a basketball court over 30 years ago, into an "enchanted forest" from the Snow White story.

As the children opened the door, they were greeted by life-size pictures of the Seven Dwarfs under the basketball hoop. On one long wall, I painted the forest, complete with Snow White's castle, a wishing well and the dwarfs' house. The other wall depicted our family tree, with 8-by-10 portraits of each family member and ancestors dating as far back as I could find. I also made 14 quilts, one for each grandchild.

The older children slept the night in the barn, accompanied by an adult, telling ghost stories and getting to know each other. During the day, they played basketball, swung on a rope from the rafters and had a talent show.

Pictures of this barn will bring a flood of pleasant memories to all of us for many years to come. —*Idora Meier*
Crescent City, California

One Halloween, my girlfriend and I decided to have a "ghost house" in our hayloft. We had a hard time deciding what to do, since we didn't have a lot of the things kids have now. Finally, we put a sign on the barn that read, "Come See the Lost Soul", then wrapped ourselves in sheets and did a lot of moaning, crying and screaming.

When the other kids climbed up to the loft, we dragged them over to a big box covered with a cloth, jerked the cloth away and urged them to peek inside. The "lost soul" inside was the *sole* from an old shoe! —*Leona Lamon, Adamsville, Alabama*

My family's barn was a great place to bring friends and cousins. Dad rigged up a swing that took us from one end of the haymow to the other. We could swing so far that we could touch the walls on either side. If I shut my eyes and remember, I still get butterflies in my stomach!

During the day, when the animals were turned out, the barn was a perfect place to play cowboys and Indians or cops and robbers. The cracks in the horse stalls were just right for spying on the "enemy" or firing an imaginary arrow or bullet. We never had to worry about getting slivers because the horses had worn the wood shiny and smooth.

As I enter the barn now, it seems much smaller than it did 50-plus years ago. But the memories of a happy childhood are still there, and the feelings are as comforting. Our barn, my childhood retreat, is still a delightful place. —*Ellen Weeks, Omaha, Nebraska*

When there wasn't much hay left in our loft, my sisters and I loved to "skate" there. The floor was slick and smooth from all the hay bales that had been moved across it, and our leather-soled shoes glided as if we were on ice.

In summer, we'd open the hayloft doors, letting the cool breeze in as we skated around. One warm day, one of my sisters "skated" right through one of those open doors! She didn't get hurt—just a little scared. I'm sure her guardian angel helped her land safely. We were more careful after that. —*Lou Voth, Muenster, Texas*

O ur barn (below) had stood lifeless for a number of years, so my husband, Orville, who's not very sentimental, thought it was pointless to make any repairs. When the roof began to leak, he thought it should be torn down before it fell down.

We couldn't afford a new roof. But as our 50th anniversary approached, the kids began to talk about a celebration. I told them I didn't want one, so they decided to put a new roof on the barn instead!

After the job was completed in late June, Orville and I went to the barn for a look. I hadn't been in there for a long time. As we spontaneously began to remove years worth of built-up chaff, I suddenly heard myself say, "Let's have a party up here." This from the woman who didn't want a celebration!

We ended up pitching four truckloads of chaff out of the loft. The floor was in surprisingly good shape but needed lots of sweeping. Our son-in-law manned an electric floor sander. It was a lot of work, and while we enjoyed it for the most part, I sometimes thought we were crazy!

Before the party, two of our daughters strung Christmas lights all around the loft and created a giant lighted "50" on one wall. I began cooking, baking and freezing. We sent personal invitations to family and friends and ran an open-house invitation in the local papers, asking everyone to wear their "barn clothes".

Two hundred people came and we had a ball! Our neighbor played the accordion, then we played tapes of dance music from the 1940's—our kind of music! That night, after all the guests left, two of our granddaughters slept in the hayloft.

People are still talking about how much fun they had. So, now our barn holds all the old memories and some new ones, too. Every time I look out the window to the barn, I remember what a great time it was.

—Lula Leverenz
Windom, Minnesota

MUSICAL ENTERTAINMENT at "Barn Day '95" was supplied by Peter Czaja Jr. (above, second from right) and his family and friends. The event was held to dedicate Peter's new barn, which he and son Mike (second from left) built from scrap lumber.

M y son, Mike, and I built my barn from the ground up, mostly with scrap lumber. It took about 5 years for Mike and me to finish the project, mostly working 1 day a week and some evenings.

I also made the cupola and weather vane atop the barn. I hand-hammered the copper cow on top and cut the arrows from brass.

When the barn was finished, Mike thought we should dedicate it by inviting neighbors and friends to a picnic. We called it "Barn Day '95". My wife and I, Mike and his fiancee and several friends provided live country, gospel and bluegrass music.

The event was a great success. At night, we had a campfire and roasted marshmallows. Friends took pictures and videos and everyone had a great time. They're already talking about making it an annual event!

—Peter Czaja Jr.
Housatonic, Massachusetts

BUILT in 1937 from lumber salvaged from the first barn on the farm, the barn at left near Windom, Minnesota was the site of a 50th anniversary bash held in September 1995 for owners Orville and Lula Leverenz (inset).

We sometimes wondered why some of our cows' milk production dropped and why they looked tired when we came to milk them in the morning. Now I think I know why!

Our barn in Minnesota was one of the largest around, with room for numerous animals, including 30 milk cows and eight horses. By October 1 of each year, the hayloft was so full of winter feed that we had to crawl on our stomachs to get from one end to the other.

There were no public places to play basketball or roller-skate, but our smooth hay barn floor was ideal for both. When we began winter feeding, we'd dig straight down through the hay until we reached the floor, then sweep that area clean for roller-skating.

At first the area would be about 10 feet square, just big enough for one person. As winter progressed, we kept clearing off more floor space and enlarging our "rink". When we had enough room, we put up our "basketball hoops"—two wooden bushel baskets with the bottoms cut out.

Soon a quarter of the floor was cleared, and we had room for several neighbors to come over to play ball or skate. We played there every evening, sometimes from 8 p.m. to midnight, and all day Saturday and Sunday.

You know how much noise one person can make while walking or working upstairs in your house. Just imagine 10 to 15 people walking, running, skating on steel-wheeled skates, jumping, dribbling a basketball and hollering overhead. No wonder our cows were red-eyed and producing sour milk! —*Curtis Nelson*
Eagan, Minnesota

My parents fell in love at a barn dance when he was a high school basketball player and she was a cheerleader. In 1936—15 years of marriage and four children later—they became the owners of that barn.

The following spring, neighbors and carpenters helped our family clean and paint the 34- by 100-foot barn. Then we built a bandstand, a concession stand and a new hayloft floor. On June 25, 1937, we hosted a dance there, and "Smith's Barn" was born.

For the next 21 years, dances at Smith's Barn featured lots of good music. There were Big Bands, battles of the bands and all-girl orchestras. The performers came from surrounding states—some from as far away as Chicago and Omaha.

During World War II, all men and women in uniform were admitted free. Graduating high school seniors got free tickets, too.

My parents belonged to the National Ballroom Association, but Smith's Barn was also home to balloon dances, wedding dances, jitterbug contests and "sunrise dances". Many couples met and fell in love there, and some returned for their wedding dances. I had *my* wedding dance at the barn in 1952.

If the barn were human, I know it would've cried the night of the last dance. But it was getting older, and so were its owners. In 21 years, they'd had to install three new floors; all those dancing feet had worn them out!

One summer afternoon, a terrible storm battered the barn, leaving it badly damaged. The barn put up a gallant fight, but it was no longer safe for dances.

It still stands today, overlooking the land. Even now, people drive by and take pictures of it, remembering the happy days when it was filled with music. —*Virginia Even*
Cogswell, North Dakota

WELL-KNOWN Smith's Barn in North Dakota (below, in photo shared by Corinne Hanson of Forman) hosted numerous dances. Below left, owner Virginia Even proudly displays memorabilia from 21 years of entertainment. Inset shows Virginia doing chores in front of the barn in the 1940's.

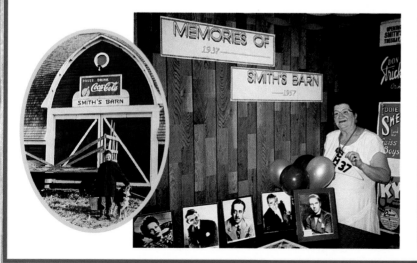

Chapter 9

Loners
And
Leaners

Bent—But Not Broken!

Folks just can't get enough of old barns. While there's admittedly something a little bit sad about these weathered and weary buildings, their quiet grace and stamina also offer hope and strength.

Standing fast against the elements, these silent sentinels symbolize the enduring strength and resilience of farmers and ranchers. They radiate a sense of history and character—staunch, tough, unyielding. More than just wood and nails, they evoke a certain mystique that's impossible to resist.

This book would be incomplete without celebrating these defiantly proud structures.

"IN 1911, my husband's great-grandfather built this barn (above) along the central Gulf Coast of Texas," relates Susan Foester, who lives close by in Port Lavaca. "It has survived numerous floods and hurricanes and is the only remaining barn of its type in Calhoun County." Susan and her husband plan to restore the barn for use in their ranching business.

FRANCIS WELCHECK built this barn for livestock around 1920 near Pineville, Louisiana. But his family lived in it, too, until they were able to build a house. The large loft was sometimes used for community activities and dances, says neighbor Alison Poston. The barn is now owned by Mr. and Mrs. William Ortego.

STATELY red barn still proudly overlooks fields near Monroe, Washington, notes Ernest Creelman of Bremerton, who snapped the photo.

VISITING her father's old barn for the first time in over 60 years brought back some vivid memories for Dru Thompson.

When I woke on the day I was to visit Daddy's old barn, I felt as though morning was dawning for the first time. I was going back to the uncomplicated world of my youth. My heart was full and my spirit soared.

How long had it been since I'd been there? Sixty years? Impossible!

Now it was just a shell, a skeleton. The proud red paint was gone. The weather-beaten roof was almost blown away. Rusty hinges held a sagging door that seemed to have closed for the last time. Could this be the magic door that used to open up to such intriguing things?

I opened it and stepped into the past and all those things seemed to return. Holstein cows filed into mangers, contentedly chewing their cuds and switching their tails at troublesome flies. Startled chickens scampered noisily for fallen grain. "Old Red", our prize rooster, flapped and stretched on a high rafter.

The weathered walls spoke to me. A squeaky board punctuated children's delighted shrieks as they jumped from the loft ladder to a soft bed of hay. Milk splattered against tin buckets, and I felt a sudden warmth on my face as my brother aimed a spray of "Bossy's" milk at me.

Was the old barn playing tricks on me? No, I could *smell* the new-mown hay and the leather harnesses on the workhorses that pulled creaking wagons from the fields. I heard the "swoosh" of hay as "Old Blue" tripped the load.

As daylight began to fade, I spotted him. Could it really be him? Yes, there he was! It was Daddy, in his plaid shirt and faded overalls, his straw hat stained with sweat, his face browned and etched by weather and time. A smile tugged at the lines on his face.

My tears ran freely as he stretched out his work-worn hand. His presence was overwhelming. But where else would I expect to see Daddy? His heritage, his faith, his pride and traditions are forever etched upon these walls.

As his image faded, I felt his love for me. Yes, Daddy's barn was now just a shell. But it holds treasures—my heritage, my memories and a love everlasting. These shall always be mine.

—*Dru Thompson, Sandy, Utah*

LONE BARN and windmill stand amid a wheat field on the Castleberry Ranch, 7 miles northwest of Ekalaka, Montana. Keith Wittenhagen of Miles City shared the photo.

UNIQUE Matt Larson Barn—the largest round barn in Minnesota—still stands defiantly in Big Stone County, northeast of Clinton. The barn, built in 1912-13, is 87 feet in diameter and 65 feet tall at the top of the cupola. As was common in those days, more expensive rock—in this case, purple granite—was used on the foundation side facing the road, while common fieldstone was used on the rear. Matt was known as the "King of the Swedes" because he often lent money to pay for other Swedes' passage to America. Son Alan and his wife, Lucille, still live on the farm. Connie Cleaver of Nampa, Idaho shared the photo.

MOODY SCENE was snapped by Christina May on her way to an Oregon ghost town. "During the Gold Rush in 1862, Auburn was the largest town in eastern Oregon, with 5,000 people, but there's not much left in the area now," says Christina, of Pendleton. "The dark clouds hanging over this barn made the scene seem even lonelier."

"WHEN I was growing up in northeastern Minnesota, this barn was always a welcoming sight as we drove along the north shore of Lake Superior," writes Terri Lillback of Niagara Falls, New York.

STRIKING but lonely barn is located on Highway 20 East near Sisters, Oregon. Richard Gann of Eaton, Colorado provided the photo.

153

aaaah...aaaah...aah...choo!

WHENEVER they'd drive along Highway 25 south of Brainerd, Minnesota, Marge and Don Rieck would watch for "the green barn". "For years, it looked like the barn could fall at any moment, so one winter day, we stopped and took some photos," recalls Marge, of Pequot Lakes, Minnesota. "The following summer, we found the barn had collapsed. We took one more picture and said a final good-bye."

CATTLE on the Bedford Farm near Jefferson, New York are now housed in a modern barn, but this three-story beauty (right) still stands as a reminder of the farm's history. Built in 1902, it measures 50 feet from the floor to the ridgepole, and the cupola is 10 feet tall. "The barn shifts with the wind, especially up in the cupola—but that's probably what has kept it standing," says Doris Bedford, who grew up on this farm.

"MY DAD'S BARN, built in 1901 at Tomah, Wisconsin, is still standing," writes Judith Asen, also of Tomah. "It's held up well over all these years."

"IMAGINE the stories this barn near Freeland in Saginaw County, Michigan could tell," muses Patty Reynolds of Saginaw. "Even with snow on top, it looks like an inviting place to be."

"THE CULVER BARN was built by my cousin's grandfather west of King City, Missouri," comments Darrel Colville of O'Fallon. "He piped in water for his mules from a large spring some distance away. A float controlled the water flow, and the animals soon learned to push the float down to get water. But there was no water piped to the farmhouse!"

ONCE-MAJESTIC BARN near Sequim, Washington was already beyond saving when Laurie Gasnick's family bought the property. "Though it had a beautiful cathedral ceiling and it was obvious that a lot of love and thought went into the construction, it just couldn't be renovated," Laurie remembers. "We were sad to see it in such a state of disrepair, and even sadder that we had to tear it down. Another piece of history has disappeared."

"THIS BARN, built in 1914, sits in the middle of a large hay and grain field 8 miles east of Sheridan, Wyoming," writes neighbor Alvin Neard. "It's just about to fall down now, but it's a delight for painters," Alvin says.

BARN owned by Ken and Joy Myers near Spencer, Idaho was built in 1919 by Charles Hardy, who provided holes in the upper face for the pigeons he raised in the loft. The fieldstone walls are 18 inches thick, and the second-story walls are made of rough-cut pine. It's listed on the National Register of Historic Places.

THIS BARN sits on an old 400-acre homestead in the foothills of Mount Rainier near Elbe, Washington. "It's so peaceful and quiet in this area," says Kenny Thomas of Tacoma. "I've counted as many as 200 elk feeding in the pasture behind the barn."

"MY GRANDPARENTS and great-grandparents traveled by covered wagon to Yellow Medicine County, Minnesota in 1877," writes Leona Meyer of Vesta. "Of all the buildings constructed on their 1,000-acre homestead, only this 1895 barn remains." Lorraine Bendix, a relative of Leona's, snapped the photo.

WHILE admiring the fall colors in their native Chautauqua County, New York, Dorothy Lundgren and her husband discovered this barn. "It stood all alone in an abandoned field," remembers Dorothy, of Jamestown. "It seemed to yearn for past years, when the fields were productive and fertile, and the mows were filled with fragrant hay."

WEATHERED BEAUTY was discovered by Cheryl Warren as she hiked along the Blue Ridge Parkway near Beech Mountain, North Carolina. "I grew up on a small farm in Virginia," says Cheryl, now of North Myrtle Beach, South Carolina. "Old barns like this one bring back memories of my childhood."

BEATEN BARN accompanies a silo that seems to defy gravity. Joan Brown of Marion, Iowa discovered this unusual landmark just west of Strawberry Point, Iowa on Highway 3.

"PASSERSBY can't help but notice this unusual red-roofed barn at the southwestern edge of Wolfe City, Texas," says Ronal Wheeler of nearby Bonham. "The barn is at least 75 years old and was built by J. Riley Green, a livestock auctioneer and cattle breeder. At livestock sales, the second-level balcony gave the auctioneer a good view of the crowd."

MOST of Janice Morris' memories of her family's barn in northeastern Nebraska are pleasant—but a few are downright scary. "The steep roof was a real challenge to climb," recalls Janice, who lives in Carroll. "But one day, two of my city cousins succeeded in climbing to the very peak!"

WITH "hat" askew, this old barn on Highway 10 between Anoka and Elk River, Minnesota looks dejected about its faded appearance. Connie Cleaver of Nampa, Idaho took the photo.

AS A CHILD, Cindy Baker loved visiting her grandparents' farm outside Carthage, Indiana. She recently returned to the area to see if their barn still existed. "To my surprise, it was still standing, although it sits alone now, with empty fields around it," says Cindy, of Bishop, California. "But to me, it's still filled with memories."

"I LIKE to take pictures of old barns with character," says Karen Hale of Wadesville, Indiana. "They're hard to find, but I'm always on the lookout." She photographed this barn near Mount Vernon in 1980. A few years later, a windstorm ripped off part of the roof, and the barn was torn down.

LONER in Madison County, New York was snapped by Dorothy Lehman for her daughter in New Jersey. "She loves old farms and barns and asked for a photo of this one," Dorothy explains. "The barn is only 3 or 4 miles from my home, so I pass it often—but I don't know anything about its history or the owner. It's looked like this for several years."

The barn stands empty now. The rafters sag, and there are gaping holes in the roof. The old hay is wet and decaying, like the barn itself. But this place holds a quiet, mysterious secret.

If I close my eyes, I can see the barn raising in 1901. The barn was treated with respect then. It housed everything imaginable—cows, Shetland ponies, pigs, chickens and kids.

Children always flocked to the barn for solace, learning and fun. The farmers who owned it made their living in it; the kids just enjoyed it.

The family time we shared made the barn special. We laughed and joked as we milked the cows and fed the calves and pigs. At morning milking, we felt fresh and invigorated, as if we were helping God start the new day.

The evening milking was filled with stories of the day's successes, failures and embarrassments; this was where we shared our joys, sorrows and dreams.

Now my children have grown and left the farm and neither the barn nor I stand so tall and straight anymore. Sometimes I feel the barn and I are aging in the same way. But the barn holds a secret—the secret of growing old gracefully.

It stands in the same place day after day, withstanding the elements and maintaining its dignity. It still houses a multitude of little critters, reminding me to live and work in harmony with nature and my neighbors.

It also teaches me to respect my rural heritage. It shows me that I may still be useful, providing shelter and hope to those in need. It symbolizes a life of service.

The old barn inspires me to do what many generations have done before me: To live out my days so that my life, too, may be an example of fortitude, harmony, service and quiet country dignity.
—Linda Van Dusartz, Pine Island, Minnesota

OLD BARN has taught Linda Van Dusartz the secret to growing old gracefully. She says its example has taught her much about living—and aging—with dignity.

DRAMATIC Chugach Mountains in Alaska's Matanuska Valley—famous for its huge vegetables—provide an awesome backdrop for lovely barn near Palmer, Alaska. Tom Dietrich of Cheyenne, Wyoming took the photo.

WHEN this barn was built around 1912 at Cottonwood, Idaho, it reportedly required 110 gallons of paint! Hay wagons were driven right inside the 110- by 60-foot barn to unload, says Becci Gehring of Keuterville, Idaho, whose husband's family once owned the farm. Although the barn has been unused for several years, Becci says people still stop to admire it and take photographs.

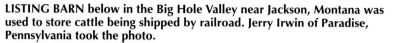

LISTING BARN below in the Big Hole Valley near Jackson, Montana was used to store cattle being shipped by railroad. Jerry Irwin of Paradise, Pennsylvania took the photo.

"I SPOTTED this interesting old barn while photographing barns in northern Maine," writes Mark Savary of Ashland. "I've photographed 85 old barns so far and have a lot more to go!"

BUILT in 1889, this weather-beaten barn is located just east of the small community of Weston in northeastern Oregon, on the western slopes of the Blue Mountains. Photo was taken by Jeffrey Torretta of La Grande.

SHARP CONTRAST between red and white paint makes the weathered barn above near Madison, Wisconsin a real eye-pleaser. Jerry Irwin of Paradise, Pennsylvania took the photo.

MOULTON BARN in Wyoming's Grand Teton National Park is one of America's most photographed barns. It was built in 1913 by Mormon homesteader Thomas Moulton and his sons Clark and Harley. Shutterbugs love its rich wood tones and the way the roof line gently mimics the Teton Range behind it. "When we built it, we didn't know we were artists," quips Clark. "We just wanted shelter for the horses." Photo was taken by Tom Dietrich of Cheyenne.

"SAGGING DAIRY BARN west of Park Rapids, Minnesota reminds us of a jack-o'-lantern left over from Halloween," write Gordon and Lois Robinson of Minneapolis. "It's gradually sinking down—and spreading out."

"OUR OLD BARN is giving up after standing proud for many years with its special cupola on top," write Myers and Marlene Rossiter of Winfield, Iowa. They don't know when the barn was built, but their records indicate the original owner bought the farm from the U.S. government in 1855—and received a patent rather than a deed. The barn is the last remaining structure from the original farmstead.

EVEN with much of one supporting stone wall missing, this barn north of Mineral Point, Wisconsin managed to stay upright—though just barely. Alfred Wells of Janesville says the barn is no longer standing.

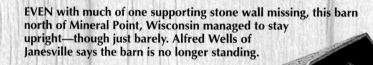

"FOR YEARS, I watched this barn near Harper, Kansas drop lower and lower," says Bob Sallee of Coffeyville. "When this picture was taken, the lowest corner was no more than 10 feet off the ground." The barn has since collapsed.

THOUGH this barn's roof is clearly giving way, it's doing so with tremendous grace, writes Kay Hendrickson of Plymouth, Michigan. She photographed this undulating wonder near Imlay City.

SWAY-BACKED BARN was spotted by Marion MacDonald of Ocala, Florida when she visited Maine for a class reunion. She was struck by the barn's sturdy appearance, despite its sagging roof and walls. The barn is located on the outskirts of Auburn.

ARMON MCDOWELL of Aloha, Oregon says he's watched this barn in neighboring Beaverton become a "sagger" over the past 35 years. "It was built in 1932 by Ed Wilson on his property off Baseline Road," Armon says.

WEAK-KNEED BARN near Delavan, Minnesota appears to be backing up, as if trying to sit down for a much-needed rest. Eugene Shumski of Fullerton, California shared the photo.

163

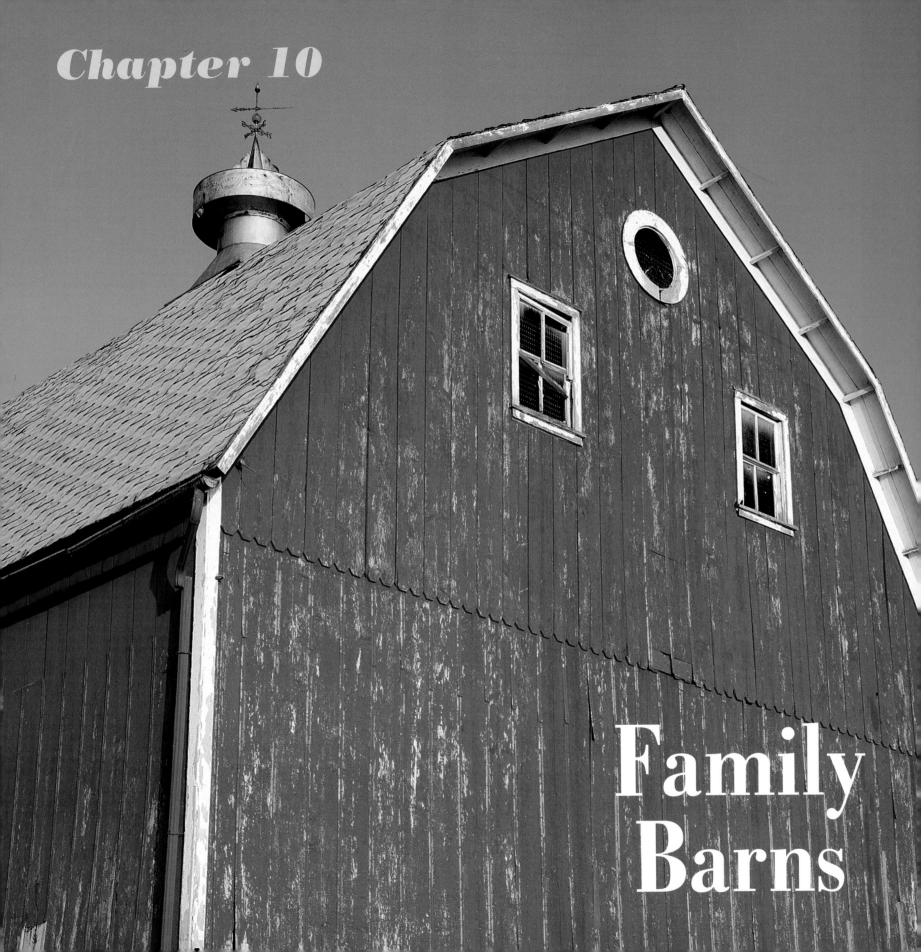

Chapter 10

Family
Barns

BEAUTIFUL STONEWORK on the Sahlbom Barn on Lake Winfield in Michigan is a fascinating example of bygone craftsmanship.

W e consider our family barn (left), on Lake Winfield near Coral, Michigan, a country work of art. My father, Rolf Sahlbom, built it with the help of a neighbor, Arlo Smith, in 1955, after the first barn burned down.

Father must have wanted something very special because he hired Bill Schenck, the best stone mason in the county, to build the first-floor wall. It was fascinating to watch Mr. Schenck carefully break the stones open and put them in the wall.

At some point during construction, Father decided he didn't want to return to the rigors of dairy farming and took a job in a factory. So the interior remains unfinished.

Today, my mother and father still live on the farm; their new house has a nice view of the barn. What lies ahead for it? There have been a lot of suggestions over the years. One thing I know for sure—you could make one terrific basketball court with all the open space inside those stone walls!

—Linda Sahlbom
Aptos, California

"AFTER our family farm's original barn was destroyed by fire, my grandfather ordered this barn by mail from Sears & Roebuck Co.," relates Dave Bishop of Atlanta, Illinois. "It came on a boxcar with everything—lumber, paint, hardware and nails. He had enough shingles left over to roof the summer kitchen. The farm, located near Groveland, has been in our family since 1867."

"MY DAD built this barn about 9 miles east of Faribault, Minnesota in 1925," writes Myrtle Durand of Faribault. "This photo was taken in 1938, when I was confirmed. I'm on the left next to my mom, Margaret; my sisters Marie and Lorraine; my dad, Roy; and my sister Violet. My brother, Ernest, was taking the picture."

"OUR FAMILY BARN, built in the 1930's out of wood from a local church, still stands about 4 miles south of Stickney, South Dakota," writes Wilma Tuttle of Mitchell. "It's special because my dad, Earl Teesdale, built most of it by himself on our family homestead—with one hand! He lost the other hand in a corn-picker accident. He could pound nails almost as fast as someone with two hands."

"THE BARN my great-grandfather, Sherman Pearsall, built west of Grand Rapids, Michigan is known as the 'Temperance Barn'," reports Phyllis Klomparens of Belmont (above right with her sister Florence). "That's because when the barn was raised in 1844, Great-Grandfather—a man of strong morals—refused to supply liquor to those helping. When told his barn couldn't be raised without serving liquor, he reportedly said, 'It shall be, or I will go without a barn!' The barn was raised, and the good people who helped departed giving three cheers—one for Great-Grandfather, one for the barn and one for the delicious meal of roast pigs."

FOR 121 YEARS, this barn was part of Kathryn Owens' family's farm, located northwest of Sequim, Washington. It boasts one of the few rounded roofs on the Olympic Peninsula. The current owners are restoring it.

A stately bow-and-truss barn on Washington's Olympic Peninsula holds many wonderful memories for my sisters, cousins and me. If I close my eyes, I can still hear the barn owls hooting and the laughter of children playing on the rope swing.

My great-grandfather built the dairy barn in 1934. A "grand initiation" was held that fall, complete with a barn dance, a hired band and tons of food.

When he stopped dairy farming, the barn became a basketball court for local children. It also was used to store tulip bulbs and boats.

My grandmother sold the barn and the last 5 acres of the farm in 1973. The current owners, Don and Penny Wolf, use the barn as a shop for their cabinet- and furniture-making business, and they're restoring it to its former glory.

The thing I'm happiest about is that their 10-year-old daughter is able to enjoy the same things I did. She has the whole second story for her "playground"!

—*Kathryn Owens*
Baxter, Minnesota

BIG STARS have made the barn built by Clarence Bradshaw's dad a local landmark outside Meadville, Pennsylvania for years. At right is Clarence's wife, Doris, in front of the barn in 1951. Clarence first heard about Doris' birth when he was 4 years old and his dad was building the barn!

My family's barn near Meadville, Pennsylvania was built in 1926 by my dad, Clyde Bradshaw, and is now owned by my nephew Joe.

The barn was always painted red with white trim and a big white star on each door. The big stars still serve as a local landmark. The barn will always be a part of me—my footprints (as a 4-year-old) are embedded for all time in the cement basement floor!

I often rode with Dad on his trips down the valley to a creek bed to retrieve sand and gravel for the concrete foundation. I really enjoyed those wagon trips with Dad and our two horses, "Polly" and "Daisy".

As we were heading for the creek bed one sunny morning in early 1926, we passed a house and Dad remarked, "They have a new baby girl in that house." At that age, I wasn't much interested in girls.

But when I returned from the service in 1945, I met the "baby", who was now a lovely young lady named Doris. To make a long story short, I've now been married to her for 50 years!

In retrospect, my dad was one of those quietly great people who helped make America the wonderful place it is. His sturdy good character is represented by the barns he and his friends built.
—Clarence Bradshaw, Erie, Pennsylvania

From 1921 to 1940, my dad, Frank Dries, rented one of his father's three farms outside Ashton, Iowa. When my grandfather died, his will required that the farms be sold and the money divided among his seven children.

Since he couldn't afford to purchase the farm on which we were living, Dad was forced to give up farming and find a new livelihood. My mother once told me that she found Dad sitting out in the barn crying on the night before we moved into town. "I don't think he ever got over the pain of losing the farm," she said.

The man who purchased our farm once told my brother, Daryl, that Dad had taken *everything* possible when we moved. "He didn't even leave behind a piece of loose lumber," the man said.

But he did leave *something* behind—a memento we didn't find for 55 years until 1995, when Daryl returned to visit the farm where he was born.

The old barn was still standing tall and proud, but the owner told Daryl he planned to tear it down soon. When Daryl asked for permission to take a few old boards as mementos, the owner told him to go ahead and help himself.

Daryl searched inside and outside the barn for boards suitable for making picture frames. He finally spotted one that seemed to be the right size and shape. When he picked the board up, he found the letters "F.D." carved into it—our dad's initials! —*Madonna Christensen, Sarasota, Florida*

YEARS after Madonna Christensen's heartbroken dad moved away from his family farm and beloved barn, her brother, Daryl, found a barn board with his dad's initials carved in it.

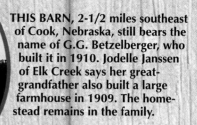

THIS BARN, 2-1/2 miles southeast of Cook, Nebraska, still bears the name of G.G. Betzelberger, who built it in 1910. Jodelle Janssen of Elk Creek says her great-grandfather also built a large farmhouse in 1909. The homestead remains in the family.

"TEN-SIDED BARN outside Strafford, Vermont was built by my great-grandfather, part of the fifth generation of our family to farm this land," observes Susan Durkee of Sharon. "My uncle, Dan Robinson, and his wife, Marge, own the farm; they're the seventh generation of our family to run the farm since it was established in the mid-1800's."

BARN near Bigfork, Montana was built in 1908 for K.B. Johnson, who sold the farm to son Ervin in 1958. Since then, Ervin has re-roofed the barn, installed a new foundation and strung cables in the loft for support. He installed steel siding after the weathered boards no longer held paint well. "The barn holds a special place in our family's memories," says Ervin's wife, Donna.

"OUR FAMILY BARN was built in the late 1800's," writes Donna Barleen of Concordia, Kansas. "This photo, taken around 1910, shows the barn and my grandfather, Herbert Lagasse, at far left with his parents, Louise and Charles Lagasse Jr. We just moved onto the farm and hope to restore the barn."

AS CHILDREN, sisters Pam Dunaway and Cheri Freeman loved to play in their barn near Danville, Indiana. The builder, one of their mother's ancestors, died shortly after the barn was finished in the 1850's. The first time the barn was used was for his funeral. Note the huge hand-hewn beams at top right.

T he barn on our farm near Danville, Indiana is truly *our* family's. The very first owner was one of our mother's ancestors. And when her family sold the farm around the turn of the century, our father's ancestors bought it!

The barn was built in the 1850's. It had a special threshing floor where the wheat was scattered for colts to run over it and thresh it.

The huge beams in the loft were hand-hewn and held together with wooden pegs. After 140 years, they still feel solid under the feet of climbing children.

As youngsters, we were always fascinated by the old stuff stored in the barn. Iron-rimmed wagon wheels and an old sleigh buried under a pile of hay provided plenty of fuel for active imaginations.

Sadly, all this is a thing of the past. There has been no livestock on the place for over 10 years, and our barn, like so many others, is falling into disrepair. That's why your book is so important.

—*Pam Dunaway and Cheri Freeman, Clayton, Indiana*

A bout 40 years ago, my father built a foundation for what he hoped would be a cow barn. Next to it, he erected a tall white-block silo and wrote my name in its concrete base.

Dad never did build the barn, but the foundation held solid and true all those years. It still seemed eager to shoulder a load at any

time. We had no idea just how long it would take!

My grandfather had purchased the farm in the 1940's. But it sat unused for decades until we inherited it in 1978. What a grand and beautiful gift! My husband, Kurt, and I were excited about turning a dream—raising our children on a farm—into reality.

After 7 years of hard work, we finally moved from town to a house on the farm. There was still no barn, but that changed eventually, thanks to our draft horse, "Kit", my husband's companion since he was 12 years old.

See, Kit "introduced" us to our neighbors by jumping the old stone fence and foraging just over the hill on our neighbor's lush lawn and beautiful flower beds.

We soon received a note in the mail from those neighbors, the Goodermotes. It read, "Dear Kurt, We would like to be good neighbors, but it's a little difficult when you don't keep your horse home!"

Kurt immediately went over and repaired what he could. And so began a friendship between Loyal Goodermote, a 77-year-old retired carpenter who could not tolerate sitting idle, and Kurt, a hardworking farrier.

As time went on, we needed a barn more and more. The existing foundation seemed so overwhelming in terms of size and cost that Kurt began leveling off another spot for a smaller building. But that big old foundation offered tantalizing possibilities to Loyal.

One day he showed us a photo of a two-story red barn. "Like

RETIRED carpenter Loyal Goodermote (above left) uses a handsaw to cut siding for the barn that took almost 40 years to build. Above right, Loyal's brother, Doug, sides the last wall. Loyal's lumber calculations were so accurate that the leftover scrap wood didn't even fill a pickup truck.

ONE of the last chores for Loyal and owner Kurt Klein (standing on board) was hanging the barn doors (above). They did a good job—the doors haven't jammed yet. Above right, the completed barn in all its splendor.

it?" he asked. "I think it's a design you can afford," he added, explaining ways we could save on lumber.

But it was still too expensive. Soon, 4 more years had passed. Every so often, Loyal would ask if we were ready to build a barn. We never were, but the blueprints seemed etched in his mind.

Finally, he just drove up one day and said, "Listen, I'm not going to live forever. Just get the supplies, and we'll build that barn!" We had learned to believe in him, so with an assist from Kurt's mother, we delivered Loyal's exacting list of supplies.

After a month of preparation, it was time to raise the barn. On the first day—July 18, 1995—the framing for the outside walls was set up. Within a few weeks, the entire frame was roofed. Then the rough-cut siding was installed. In mid-August, we left for a week of vacation.

By the time we returned, Loyal and his brother Doug had completed the siding. Loyal's lumber calculations had been so precise that the scrap from the entire project barely filled a pickup truck!

As we stood trying to comprehend their generosity, Loyal said with a shy smile, "Well, there's always time to give somebody a hand."

These days, I often watch the activity in the barn from my kitchen window. It's especially heartwarming on winter nights, with the big door open and the mellow lights showing the silhouettes of eager horses being led to their stalls.

The aroma of hay, horses and saddles and the sight of our children experiencing the pleasures of living on a farm fill my heart with deep gratitude for all who helped make the long-awaited barn-raising a reality. —*Sherri Klein, Berlin, New York*

In the 1920's, my grandparents, George and Rosa Powers, needed a larger barn for their southeastern-Kansas farm. Grandpa started cutting walnut trees along the creek, then had them cut at a sawmill about a mile away. Construction began in the winter of 1926.

The foundation work and framing went along fine. But on morning of February 10, 1926, as the men were putting on the siding, things suddenly got exciting as Grandma went into labor with her sixth child.

The doctor came quickly, but Grandma was having difficulties. Another doctor was called in. The physicians feared Grandma might need surgery.

Now, you might think that since this was Grandpa's sixth child, he'd be used to all the excitement. But Grandma had never had a difficult delivery before—and she'd never had a baby while Grandpa was in the middle of building a barn!

Grandpa became a little panicky. All day long, he ran out to the barn site to check on the work crew, then ran back into the house to check on Grandma.

When the baby, a boy, was finally born late that afternoon, Grandpa must have breathed a huge sigh of relief. Grandma was fine, the baby

FLOYD POWERS stands outside the barn that was raised the day he was born in 1926. The farm is now owned by Floyd's son.

was healthy and the barn was still being built. They named the baby Floyd.

That baby grew up to be my father. As a child, he loved playing in the barn with his brothers, making forts and tunnels in the loose hay.

Whenever they got too rowdy in the house, Grandma would say, "You boys go play in the barn!" They were always happy to oblige.

The barn looks old and worn now, patched here and there with tin. But it still holds great meaning for me and my family. When I became a parent, I was so happy to be able to take *my* children to the barn to play. They romped in the hay, swung on the rope and looked for baby kittens—just as I used to do when I was their age.

My dad and the barn are both worn by time, but they're so much alike—warm and strong, with a sense of purpose. Seeing the two of them together brings a flood of memories to all of us who know and love them both.

—Lesa Stafford
Thayer, Kansas

"THE 'GOOD ENOUGH' SIGN on our barn, located 2-1/2 miles southeast of Peru, Indiana, makes it a local landmark," relates Joey Kubesch of Peru. "Family legend has it that my grandfather, J. Omar Cole, visited the farm home with his bride, Josephine, and asked her how she liked it. 'Omar, it's good enough for me!' was her reply."

KIMMELL BARN was built by Mary Ann Freese's grandfather and his neighbors shortly after the turn of the century near Covington, Oklahoma. The barn was constructed with locally quarried sandstone and lumber from the St. Louis World's Fair. Mary Ann and husband Orval still use the barn for hay and cattle.

Frontier ingenuity played a role in the construction of our barn—and so did the World's Fair in St. Louis!

My wife's grandfather, Sam Kimmell, settled here in 1893. Around the turn of the century, he was ready to build a barn, but building materials were scarce and expensive.

Sam decided to use sandstone for the exterior walls and foundation and hauled rocks from a quarry a mile from the farm. Some of the stones were so large that the horses had to haul them one at a time.

Finding lumber was more difficult. A neighbor, Scott Craig, needed lumber for a house, so he and Sam pooled their resources and bought some from the World's Fair in St. Louis. When the fair's Louisiana Purchase Exposition was dismantled in 1904, they had the wood shipped by rail as far as Perry, then loaded it onto horse-drawn wagons for the final leg home.

With help from his neighbors, Sam finished the 40- by 60-foot barn in 1906. The year was chiseled in stone above one of the doorways. The barn reflected Sam's Pennsylvania Dutch heritage, with simple lines softened by a louvered square cupola, round windows in both gables and weather vane lightning rods with glass globes.

My wife, Mary Ann, and I purchased the farm from her parents in 1974. It was listed on the National Register of Historic Places in 1982. Except for some tornado damage in 1922, the barn has weathered the years well and remains in good shape today.

—Orval Freese, Covington, Oklahoma

171

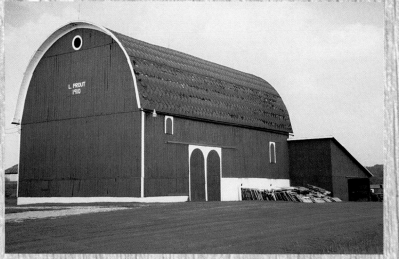

"MY husband's grandfather built this barn near Rosebush, Michigan with the help of neighbors, family, friends and two hired carpenters in 1914," writes Christine Prout. "He'd built his house in 1910, so he put that date on the barn." Christine says her in-laws, Lloyd and Lillian Prout, still live on the farm.

GYPSUM ROCKS mined from nearby creek banks formed the foundation for the Oklahoma barn below, completed in 1910. "The rocks were hauled to the site by horse-drawn sleds, and then my grandpa, Charles Raab, planed and shaped the rocks with special tools," says Therese Raab-Smith of Custer City. "It took him and my great-great-uncle, Frank Raab, one whole fall and winter to build the barn."

"THIS BARN was built by my great-grandfather, Joseph Gore Evans, and has been in our family for 115 years," says Michelle Holdsworth of Nashport, Ohio. "I've heard my grandfather tell many times how his father hewed the logs into beams and used wooden pegs to hold them together. My parents own the barn now. Although we can't afford to restore it to its former glory, it remains a comforting place to us."

"OUR OLD LOG BARN must be at least 150 years old," writes Lois Bode of Clendenin, West Virginia. Her father purchased the farm in the early 1920's. "On Sunday afternoons, we children would sit in the open haymow window with our feet hanging outside, while we cracked hickory nuts to eat," Lois remembers. "In fall, my father put baskets of apples under the hay so they'd keep for winter."

The barn on our family's Ohio farm is very special to all of us, because it links us strongly to our past. My great-great-grandfather, Johann Albers, homesteaded in Auglaize County, Ohio. He and his young family arrived from Germany around 1833, living in a temporary shelter until they finished their log house.

Johann's son, Anton, built a brick house in 1872, then started work on the barn. It took a year to hew the logs and another year to hand-split all the roof shingles. The building is held together with handcrafted pins made of tough hickory. Construction was finished around 1877.

As a child, I remember looking up in awe at the hand-hewn log that supports the bank barn's upper floor. Today, my brother Anthony, who runs the farm with his son Michael, uses the barn to shelter calves.

My siblings and I enjoyed growing up on a farm. We have a sense of who we are because of our connection to those who went before. The farm gave us firm roots and a family history we're proud of. In a very real way, we are still touched by those ancestors who came before us and built this barn.

—*Regina Albers, Adrian, Michigan*

HANDSOME BANK BARN was built on the Johann Albers homestead in Auglaize County, Ohio around 1877. Johann's descendants still run the dairy farm. Inset photo shows hand-hewn log used inside the barn.

"MY GRANDFATHER, Curtis Strang, built this barn of lava rock, cement and wood near Sterling, Idaho in 1908," says Anne Locke of Santa Clara, California. **"The construction was no easy task. All the supplies had to be hauled from American Falls, about 30 miles through sagebrush."** Photo above shows the barn in 1913. In its early days, it was used as lodging for newcomers to the area. It's now a cabinetmaking shop, at left.

BUILT IN THE EARLY 1860's at Hadley, Pennsylvania, barn at left has been in Lu Ann Shreckengosh's family for five generations and was home to dairy cows for many years," writes neighbor Shirley Kather.

"WHEN my great-grandfather's barn (right) was completed in 1891, it sported a 15-ton slate roof," says Charles McDonald of Inwood, West Virginia. "Most of the lumber came from the farm, as did the foundation stones. The barn, which I now own, is one of the largest in the area—85 feet long and 65 feet wide."

STRIKING BARN on the Spring River near Verona, Missouri is on part of the century farm that's been in Tom Davenport's family since 1834. "The barn was built in the 1890's, using timber cut from the land," says Tom, of Marshfield. "The framework was ax-hewn and no nails were used, except for the siding and roof."

NORMA HETRICK'S grandparents raised 15,000 chickens in this barn from 1900 to 1936, when her parents bought the property. "A semi truck came by once a week to pick up the eggs," says Norma of Fremont, Ohio. The barn, which features 2-foot-thick walls, is about 150 years old.

SERIES OF ROPES held a 32-foot fireman's ladder in place as Ernest Swanson's family repainted his barn. Daughter Judy Piper of Eagle Nest, New Mexico says the project took 15 gallons of paint. Ernest and his father had the barn built near Angel Fire, New Mexico in 1936 at a cost of $4,000 to $5,000. Photo shows Ernest (in cowboy hat) and son-in-law Burk at work.

T he words "Meadow Home Dairy" are long faded from the roof, but most people around Eureka, Montana still recognize our wonderful barn. With the Rocky Mountains as a backdrop and surrounded by rolling green hills, the barn is a favorite subject for local photographers.

My husband Dale's grandfather, Arthur Purdy, commissioned the building in 1923. The contractor got most of the lumber from a dance hall 3 miles away and labeled each piece to indicate where it would go. When all the boards were numbered, Arthur rounded up a crew to assemble the barn.

It measured 40 by 72 feet and originally was used to hand-milk 20 to 30 cows. (Raw milk sold for a nickel a quart at the time.) The large loft could hold 80 tons of loose hay, and it also had a 20-ton grain bin.

When the loft was no longer needed for hay storage, it became a giant play area for numerous friends and relatives. A swing was installed, old saddles were placed on hay bales for youngsters to "ride" and a hoop turned the space into a great basketball court.

Our Baptist Church held Christmas Eve services in the barn for 3 consecutive years in the 1980's. Unfortunately, this impressive service became *too* popular and safety concerns forced us to discontinue it.
—*Marie Purdy, Eureka, Montana*

JESSIE AND LYNN PURDY posed in 1944 by the well-known barn built by his father, Arthur, outside Eureka, Montana in 1923. Their son, Dale, and wife Marie have worked, played and worshiped with family members and friends in the same barn (below, as it appears today). Their son, Joe, is now buying the farm; he'll be the fourth generation to run it. Dale's niece Janice Purdy of Coeur d'Alene, Idaho provided the photos.

175

BECKY CROSLIN remembers watching her father build this tobacco barn in Logan County, near Lewisburg, Kentucky, about 60 years ago. The "home place" is now owned by her brother, Bobby Brown. "Like all his other buildings, he keeps it shored up, painted and well cared for," says Becky, of Dallas, Texas.

USING TIMBER cut from trees on his farm, Wayne Gleason's grandfather built this barn near Marshall, Michigan in 1904. Construction costs totaled $750; 51 years later, the family spent that amount just for a new roof! "I was born on this farm, and for 81 years, it was the only address I ever had," Wayne says. The farm is now owned by his niece and her husband.

PICTURESQUE BANK BARN in Clearfield County, Pennsylvania was built in the form of a double log cabin. "My great-grand-father, John Shirey, cleared the land, then hand-hewed white pine logs to build the barn around 1860," says Nancy Hopkins of State College. "Downstairs, one ceiling beam runs the entire 70-foot length of the structure. My mother, Myrtle, still lives on the farm."

NICHOLAS HOUSE paid $250 to have this barn built south of Merrill, Michigan in 1917. He provided all the wood and made the cement blocks for the interior walls. Nicholas' daughter, Marian Lauer of Ithaca, Michigan, shared the photo.

A ROVING PHOTOGRAPHER took the photo below of Shirley Sutton's family barn near Garnavillo, Iowa in 1906. "Notice the cupolas—the carpenter was very artistic," Shirley says. "A man named William Splise is holding the dapple gray horse and a hired hand is holding the black horse. The little boy was the photographer's son." Photo at left shows the barn as it appears today.

UNIQUE WELSH BARN outside Le Raysville, Pennsylvania is owned by John and Jane Upham. It was built in 1889 for Levi Upham, John's great-grandfather. So far, five generations of Uphams have lived and worked on the farm—and a sixth is waiting in the wings!

Welsh barns like ours (above), which was built for my great-grandfather, Levi Upham, in 1889, are unique to our area. A carpenter named Morgan Thomas built 10 or 11 barns of this style for farmers in the Welsh settlement of Neath, in eastern Bradford and western Susquehanna Counties.

Ours was the first. The barns feature three stories—a stable for cattle on the basement level, a two-story-high haymow and granary on the second level and a third level—only over the granary—where horses and wagons (and later, tractors) were driven into the barn.

The third floor is the unique part. It's accessible by an earthen approach and a bridge. After wagons were pulled in, hay could be unloaded *down* into the mows—much easier than unloading it *up* into a mow, as is common with most barns.

Lumber for our barn was cut a year before the barn was built. A big flood scattered it all over a pasture, but it was gathered up and still used. Mud from the Flood of '88 is still visible on some of the roof boards!

Today, three generations of Uphams still work on our farm—my father, Richard, and my Uncle Edwin; myself and my wife, Jane; and my son, Rick, and his wife, Lisa. We're hoping that Rick and Lisa's sons, Bryce, 5, and Kent, 4, will someday be the sixth generation of Uphams to live and work on this farm.

—John Upham, Le Raysville, Pennsylvania

"THIS BARN is a treasured landmark on our family's farm," relates Mrs. Glenn Mabe of Hillsville, Virginia. "My late father-in-law built it around 1905. The tin roof put on in 1935 replaced the original shingles." The farm is in Carroll County, near the Blue Ridge Mountains.

178

INTERIOR of this 70-foot-long barn hasn't been changed since Mayo Thoreson's grandfather built it near Mayville, South Dakota in 1917. "He homesteaded this place in 1879, and my father was born here," says Mayo. "Now I make my home here, in a house built in 1890. After I'm gone, the farm will be kept in the family by my son, Douglas."

"MY FATHER called this a 'pole barn' because all the framework and joists were small tree trunks with the bark still on," says Robert Gall of Independence, Missouri. The barn, built on the family farm near Turney in the 1890's, sits on large flat rocks rather than a foundation. Robert says the barn has leaned for as long as he can remember—so much, in fact, that the sliding doors had to be cut off so they could still be opened. "They still need to be trimmed every few months," he says.

"MY GRANDFATHER built these three barns in 1890, using timber he cut on the farm," writes Kenny Alstrom of Tawas City, Michigan. "We've always tried to keep the barns maintained; they could very well last another 100 years."

179

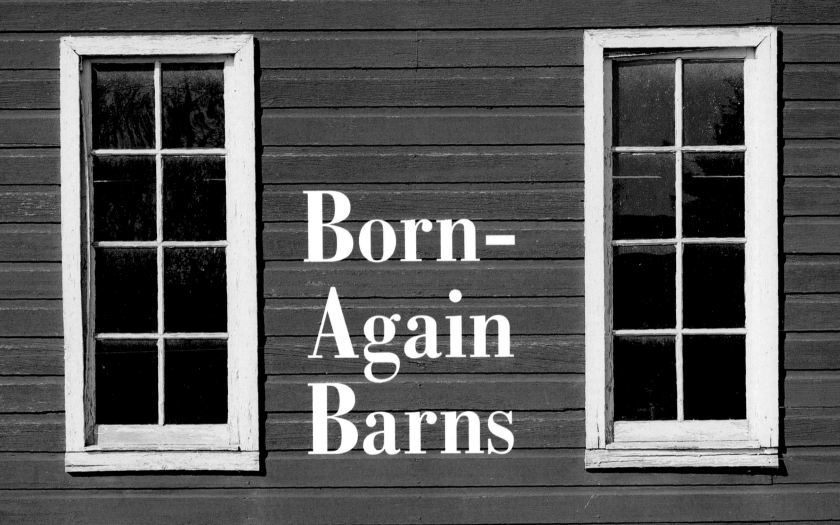

Chapter 11

Born-Again Barns

ROUND BARN on Highway 15 between Jefferson and Commerce, Georgia has housed a furniture business for over 30 years. Built in 1913, the barn was recently added to the National Register of Historic Places.

In 1939, my future father-in-law, Hugh Maley, bought a farm that included a round dairy barn built in 1913. But he found the barn expensive to maintain and allowed it to deteriorate over the years.

After World War II, I married Mr. Maley's daughter and spent 18 years working in the furniture business. One day I realized that with a little work, that old barn could be turned into a furniture store.

In 1963, I made some rough repairs, painted the barn red and fixed the ramp so we could move furniture to the second floor. There were still cows in the ground-level stalls at that time, so it wasn't unusual for a customer to look at furniture while hearing the cattle lowing down below!

Eventually the cows were moved out and we turned the ground level into more retail space. We also put a floor in the top of the central silo to make an office.

We never tried to disguise the fact that our business was in a barn. We left the interior rough and unfinished and displayed kitchen chairs by hanging them from the rafters. I retired in 1994, but the business is still run by our daughter and her husband.

—*R. Louis Turner, Jefferson, Georgia*

"OUR HISTORIC BARN was built in 1884-85 by Edward Allis of the well-known Allis-Chalmers company," writes Vernon Michel of Preston, Minnesota. "We've restored it and made it the centerpiece of our Old Barn Resort (below). The barn houses a 53-bed hostel, a restaurant, an Amish craft shop, a grocery store, a lounge and an office." The pre-restoration photo above was taken in 1988.

181

"MY GRANDFATHER, William Toogood, built this barn (at right) in the early 1860's," writes John Toogood of Rochester, Minnesota. "The farm's original grant papers were signed by Abraham Lincoln." All the stones were hand-quarried, then hauled to the building site on ox-drawn stone boats. The remodeled barn now is a dentist's office. The black-and-white photo below was taken in 1928.

WITH FUNDING from a $1.2 million bond issue, this one-time dairy barn became the Dorey Recreation Center, a public facility in Richmond, Virginia. The farmland that once surrounded the building is now a park. Neighbor Audrey Bowers provided the photo.

HOPE'S BRIDAL BOUTIQUE has been located in this sizable barn 30 miles west of Cedar Rapids, Iowa since 1972. Joan Brown of Marion, Iowa provided the photo. "The barn sat empty for several years before it became a bridal shop," Joan recalls. "It was a dairy farm at one time."

LITERALLY a bank barn, the brick beauty below was built in the 1920's by H.F. Walter, a hog farmer from Bendena, Kansas. Local resident Jesse Barrow says the barn was partially restored for business use in the 1970's and renovated further after the Bendena State Bank bought it in 1988. "Feel free to visit," Jesse advises. "The owners and employees enjoy sharing their bank with guests and customers."

INVITING DANCE FLOOR is a big draw at Larson's Barn near McGregor, Minnesota. Owners Lowell and Alice Larson host barn dances every Saturday from April to October. "The popularity of these dances has forced them to add on to the barn three times," says Margaret Weston of Palisade. "It now seats over 300. The bands have brought dancing pleasure to people from many miles around."

LAST ROUND BARN in Wayne County, Iowa was bought by a community group and converted into the International Center for Rural Culture and Art. Group member Ethel Jordan of Corydon provided the photo of the refurbished barn (above right). Linda Bryan of Allerton shared the photo above showing how the barn looked before work began.

M y siblings and I never gave much thought to growing up on a farm with a round barn—it just meant that everyone knew where we lived!

Our parents, Beulah and Cecil Bryan, moved to this farm just outside Allerton, Iowa in 1936. It was a wonderful place for five children to grow up. The barn remained part of our family until 1963, when my brother and his wife sold the farm to another couple.

When those owners retired in 1983, the barn was abandoned. As time passed, all the surrounding structures—the grand old farmhouse, the cattle barn, the corncribs and the long-silent henhouses—were torn down, one by one. The old barn stood alone, withering in the natural elements.

In 1991, a group of local citizens learned the barn was going to be bulldozed. They obtained a loan, bought the 93-acre farm site and converted the barn into a rural arts and cultural center. The last round barn in Wayne County, Iowa had been saved.

The barn now has a new cedar-shingle roof, a cupola designed to look like the original, and new windows and doors. Native grasses and flowers have been reintroduced to the grounds. The goal is to eventually return much of the site to native prairie.

The historic Spring Branch Church also was moved to the grounds. A one-room schoolhouse has been donated for the site, and there are hopes to someday obtain a turn-of-the-century farmhouse as well.

The barn is open for tours and hosts art workshops, art and quilt shows, family reunions and parties. After the haymow floor is reinforced, barn dances will be held there, too.

As you stand at the barn under the vast sky, you can see what this land must have looked like when it was still open prairie. It helps us to understand where Iowa people have been—and reminds us that as we move forward, the past is always with us.

—*Marcilee Yeager, Newton, Iowa*

REFURBISHED MULE BARN serves as the clubhouse at the Clark family's Brickyard Plantation golf course in Americus, Georgia. "We built the 27-hole course on a farm our grandfather owned at the turn of the century," explains Mrs. W.N. Clark. "The renovated mule barn houses our pro shop and offices, with a conference room in the old hayloft."

WHEN William Taylor needed to relocate his medical practice, his wife suggested remodeling their barn in Lewisburg, Tennessee. "Within a year, the barn had been reborn as Taylor Clinic," recalls daughter Julie Plott of Murfreesboro. "Dad practiced medicine there for 17 years until he retired in 1989. Now it's being spruced up for two local doctors starting a new practice."

ARCHED-ROOF BARN at Albert Lea, Minnesota was "born again"—twice! Loren Osman of Shorewood, Wisconsin, who took the photo, says the dairy barn sat empty for a time, then provided classroom space for a small college. When the school closed, the Normanna Lodge of the Sons of Norway bought the barn. "It was in terrible shape," recalls lodge historian Beverly Wessels. "But we're fortunate to have members who know how to get things done."

A MULE BARN built in 1860 is now part of Weston Red Barn Farm, an educational facility west of Tracy, Missouri. The barn houses sheep, goats, horses and a country store. The working farmstead is open to the public from early spring through late fall. "My son owns this farm, which is toured by thousands of schoolchildren every year," writes Mrs. Robert Frey of Overland Park, Kansas.

CONTRACTORS and congregation members worked together to turn the barn at right in Brookfield, Wisconsin into Countryside Christian Church (above). "We moved into our new 'home' in September 1994," says parishioner Robin Zdroik of West Allis.

"THIS 100-year-old barn was converted into a church near Loveland, Ohio," writes LeRoy Schultz of Morgantown, West Virginia. "There's also a day-care center for children in the basement."

IT TOOK a small Pennsylvania congregation nearly 6 years to turn an old barn (below) into a church (below right). Members of the Pocono Church of Christ bought the barn in Stroudsville in 1984, and spent about 2 years just clearing out old hay and other debris. Member Gary Turner of Blue Bell says the renovation got a big boost when a church in Tennessee sent a team to help with wiring, plumbing, concrete work and other tasks. In 1990, 250 members and guests attended the new church's dedication ceremony.

"CONVERTED BARN sits on a country road in Oceana County, Michigan," says Carmen Carter of Fremont. "It was first a barn, then a restaurant and now it's a church."

OLD BARN above was built in the 1930's. It's hard to believe it now looks like this inside (right) and this outside (above right). The renovation took 5 years to complete.

If we build it, they will come. With that belief in mind, Southern Baptist missionaries Jerry and Janice Jones began the unusual but rewarding task of transforming an old barn near Ellington, New York into a beautiful country church.

You see, when the Joneses first were called to this small community near the Pennsylvania border, there was no Southern Baptist church. For years, they used an old hall on the town square as a substitute.

But by 1989, the congregation had outgrown the small building, prompting a search for new quarters. And like in the movie *Field of Dreams*, the Joneses found what they were looking for in a cornfield. But rather than a ball field, they found an old barn.

Known as the Rexford Dance Hall, the barn had stanchions for cows in the lower section and a hardwood dance floor up above. With a lot of faith, vision and perseverance over the course of 5 years, Jerry and Janice's barn-to-church dream became the dream of others.

Today, children's footsteps have replaced hoofbeats in the lower section, where Sunday school classes are held. And hymns have replaced dance tunes on the top floor.

More than 30 work crews from the Georgia Baptist Association helped with the renovation, known as "Operation Clean Sweep". Workers also came from Alabama, Virginia and the Carolinas. Extra outbuildings and silos were razed, leaving only the gambrel-roofed main barn intact.

"For a while, our motto was, 'Join the Baptist Church and learn a trade,'" quips Jerry. "This work wasn't done by professionals—just by people who pitched in and worked.

"It sure helped that the building was square and level—probably more so than modern buildings. It was not more than an inch off anywhere."

The upper-level sanctuary is even handicapped-accessible, thanks to an elevator donated from a private home in Virginia. Jerry and a co-worker took videos of the elevator in action, then used the footage to figure out how to install the elevator in the church.

And people have come. The Baptist Church of Ellington now boasts 130 members. As one member, Virginia Waldron, noted in a poem:

> *God's able to use anyone anywhere*
> *If we're faithful and hear His plea.*
> *See this barn that once housed cattle*
> *But now spiritually feeds you and me.*
> —Anne Swanson, Russell, Pennsylvania

BEFORE George and Bette Cook called this structure (below) home, it was a barn housing a recycling center for the village of Litchfield, Connecticut. They bought the barn in April 1978 and moved in the following October. "I wasn't done working on it then by any means, but we made it through that winter okay," George recalls. "Everyone thought I was crazy, but as the work progressed, they all changed their minds."

"THIS HOUSE was created from an old barn," relates Barbara Iversen of Verona, Missouri. "The windows at the top of the silo provide a picturesque view of the area."

"AS A SMALL CHILD, I used to dream about living in this house (above), which I'd see whenever I visited my great-grandmother, who lived across the street," writes Caroline Manis of Woodston, Kansas. "It used to be a barn; back in 1924, a young man named Art Koontz moved it into Woodston. The initials of all his children are neatly etched in the cement cistern on the north side of the house. In 1970, my dream came true when we moved into this wonderful house."

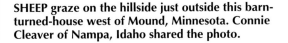

SHEEP graze on the hillside just outside this barn-turned-house west of Mound, Minnesota. Connie Cleaver of Nampa, Idaho shared the photo.

WHEN Mr. and Mrs. Richard Roesch became "empty nesters", they wanted a smaller house but didn't want to leave their farm near Springfield, Illinois. Turning their 1937 barn (above) into a house (left) was the perfect solution. Their living quarters are on the first level, with two-story guest quarters in the loft. An attached cattle shed was converted into a three-car garage. "Despite all the work and expense, we feel it was worth it," Mrs. Roesch says. "We've had countless visitors, and it's the preferred gathering place for our family. We're very proud of it."

BARN near Mora, Minnesota was purchased by Dennis Strom in 1955, 2 years after it was built. In 1979, he began converting the space into four apartments. "At first, we were concerned that no one would want to live in a barn apartment 8 miles from town," Dennis recalls. "Now we have a waiting list!"

WHEN Connie and Hershal Shaffer bought a farm in Union County, Kentucky in 1989, they considered it a challenge to convert the barn (inset) into a house (left). "With Hershal's experience as a carpenter, and the help of some Amish neighbors and others, we were able to move in after 2 years of hard work," Connie says proudly. "We love telling people we live in a barn!"

IN THE 1870'S, this barn was undoubtedly a source of great pride to the dairyman who built it. Now it contains small stores in New Milford, Connecticut.

BUILT in 1890, this general-purpose barn has been converted into an antique store outside Mount Pleasant, Pennsylvania.

OCTAGONAL BARN near Washington, Pennsylvania was built in 1888 and originally housed livestock (above). Today, it houses the offices of an architectural firm (below).

CIRCUS HOUSE RESTAURANT at Rockland, New York is aptly named—it originally sheltered elephants for the Barnum & Bailey Circus! The six-sided barn was built in 1920 and converted into an eatery in the 1970's.

Once again, we decided to dip into LeRoy Schultz's huge collection of barn photos, which include many reborn barns. As explained on page 119, LeRoy, from Morgantown, West Virginia, has snapped more than *15,000 photos* of barns across the country during the last 20 years. In doing so, he's traveled over 1 million miles through almost every state in the continental United States and worn out five cars! His collection is as unique as the "born-again" barns on these pages.

CENTURY-OLD BARN above is now home to the tennis team at Indiana University of Pennsylvania in Indiana, Pennsylvania. The barn was built in 1891 for cattle and hay storage.

FARMSTEAD built in 1887 near Huron, Ohio now is home to retail shops.

191

IN 1938, a tornado destroyed everything on Cleve Holt's farm near Columbus, Kansas. Neighbors helped him build a new barn out of lumber salvaged from a flattened house. In 1989, the barn again became a house when his granddaughter Marva opened the Country Loft Bed-and-Breakfast (above). "My grandparents and father have passed away, but I'm sure they'd be happy if they knew what the barn is doing now," Marva says.

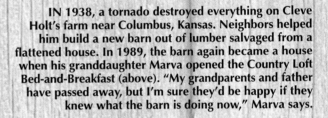

OLD BARN near Augusta, Michigan was half built when aspiring actor and director Jack Ragotzy leased it for a young summer stock company, the Village Players, in 1949. The group, now known as the Barn Theatre, celebrated its 50th anniversary in 1996, having played to nearly 2 million patrons. When Jack found the barn in 1949 (see photo at top left), it had a single light socket and one faucet inside. The troupe built a stage out of boards (above left) from the barn's top floor. Today, the spruced-up barn (left) is surrounded by a complex of other production buildings on 75 acres of groomed countryside.

USING THE WOOD from an old deteriorating barn, Tom Bruins (at left) built his own barn (above). Photo above right shows the barn's solid framework.

In the summer of 1992, I regularly mowed the lawn for an elderly woman who lived on a farm with a deteriorating barn. Part of the roof had collapsed and the large door had stood open for years.

She wanted the barn torn down for safety reasons. I needed a storage building. After some discussion and planning, we had a plan to satisfy both our needs.

Family members helped me take down the old barn, which measured 30 by 50 feet, and reshape it into a 24- by 32-foot salt-box-style barn on my land. It was challenging, painstaking work, but we finally had a barn raising on October 7, 1995.

Members of our church's men's group helped raise the walls and rafters, and the women provided food. It was a great day, even though it rained some. It was a good feeling to work together and accomplish so much in a single day.

—*Tom Bruins*
Allendale, Michigan

BUILT around 1790 near Holland, New York, this handsome barn is now part of a Christmas tree and landscape plantings farm owned by Ann Zywiczynski's family. "My husband and sons added dormers, built a new entry and clad the building with lumber from our woods," Ann explains. "Shelves and archways display local crafts. It's a quaint atmosphere, and our Christmas tree customers enjoy warming up with hot chocolate and coffee in our 'recycled barn'."

Grandpa's Old Barn

Grandpa's old barn stands high on a hill,
though weathered and old, it is sturdy still.

It creaks and groans while bracing the wind,
and just like Grandpa, it's a faithful friend.

It shares my secrets of an earlier year
and memories of youth that I still hold dear.

Memories remain of when I played oft
in the fragrant hay piled high in the loft.

Newborn calves frisked about in their stall
with nary a care or worry at all.

A litter of kittens with fur soft as silk
incessantly begged for a treat of warm milk.

A cozy haven from the cold and the rain,
my grandpa's barn, I remember again.

For a romp in the hay without a care or a darn,
I'll return up the hill to Grandpa's old barn.

—*Charles Clevenger, New Boston, Ohio*